BIBLIOGRAPHIC RECORD TARGET

AAS8341
UL FMT B RT a BL m T/C DT 07/15/88 R/DT 07/15/88 CC STAT mm E/L 1
010: : |a a 10000953
035/1: : |a (RLIN)MIUG84-B52749
035/2: : |a (CaOTULAS)160186757
040: : |a Columbia Univ. |c MiU
100:1 : |a Berkeley, Hastings.
245:00: |a Mysticism in modern mathematics, |c by Hastings Berkeley.
260: : |a London, |a New York [etc.] |b H. Frowde, |c 1910.
300/1: : |a xii, 264 p. |b diagrs. |c 23 cm.
650/1: 0: |a Mysticism
650/2: 0: |a Mathematics |x Philosophy
998: : |c KLB |s 9124

MYSTICISM IN MODERN MATHEMATICS

BY

HASTINGS BERKELEY

HENRY FROWDE
OXFORD UNIVERSITY PRESS
LONDON, EDINBURGH, NEW YORK, TORONTO
AND MELBOURNE
1910

OXFORD: HORACE HART
PRINTER TO THE UNIVERSITY

PREFACE

To the Pythagoreans belongs by common consent the distinction of having raised mathematics to the level of a science. We may well feel some astonishment when we reflect that this great achievement, which implies in the achievers the spirit of sober scientific reasoning, should nevertheless have been the work of men who were also enthusiasts and mystics. But we can find it in no way strange that the complication of these disparate tendencies should have resulted in the profession of philosophical doctrines, respecting the world-significance of numbers, which to us appear fantastic to the verge of absurdity, and which not improbably wore that appearance to some few contemporary intellects, more critical if less original and creative.

When we consider these peculiarities of mental constitution, reflect upon the nature of the intellectual matrix which brought forth mathematics as a science, and call to mind that universally, and at all times, curious superstitions and extravagant notions appear to have been associated with ideas of number, the hypothesis not unnaturally suggests itself that even now the philosophy of mathematics may not wholly have freed itself from mystical implications. Such a supposition, though it may prove to be ill-founded, can hardly be regarded as inherently improbable. Not, indeed, that there is any obvious connexion between mathematics and mysticism: the former is a subject of thought, the latter, to speak generally, is an attitude of mind, probably the effect, or the concomitant, of temperamental disposition. But it may be that all ratiocinative processes, no matter what the subject, in which the current and continual substitution of symbols (of any kind) for concepts is a prime condition of the effective conduct of the process, are provocative of that attitude of mind. I believe this to be in general the case, though with great differences between individuals; and I incline to think that not infrequently the more gifted is the individual for the prosecution of purely symbolic trains of thought the greater is the provocation to this mental attitude. Ability to resist this tendency to lapse into the mystical must, however, increase more and more with the growing insight into the nature of our mental processes which is afforded us by the progressive development of psychological investigation. The recent emergence in the domain of philosophy of what is known as Pragmatism is clearly a result of this progressive development, and Pragmatism may not inaptly be characterized as a methodical and determined attempt to rid philosophy of mysticism.

Of the two mathematical doctrines discussed respectively in Part II and Part III of this book, the first—that of Imaginary Quantity in Algebra and Imaginary Loci in Geometry—seems to have attracted but little attention from the philosopher pure and simple. Discussion of the principles which underlie this development of mathematical expression has been confined almost exclusively within the inner circle of mathematicians interested in the philosophy of their science. But wellnigh up to the close of last century the explanations of this development—especially with regard to Imaginary Quantity—were almost purely mystical: they really explained nothing, threw no light on the motives and processes of thought which obscurely prompted it, and thus left it as much of an enigma as ever. In recent years, however, a real endeavour has been made to unravel this long-standing difficulty in the development of algebraic symbolism; it is even averred that we have now at last a completely adequate explanation of it. The claim is, in other words, that the doctrine has been purged of its mystical implications. But, as I hold, the claim outruns the performance: the purgation is not thorough, and Part II of the present work is in the main an attempt to make it thorough.

Very different were the circumstances which attended the growth of the doctrine discussed in Part III. Metageometry, or Pangeometry, or the theory of non-Euclidean spaces and geometries, was mainly developed in the second half of the nineteenth century. It gave rise to a controversy in which the opposition to the doctrine proceeded almost wholly from the philosophers, especially from the Kantians, the great body of mathematicians meanwhile standing aloof from the conflict and apparently taking little or no interest in it. That the philosophers were worsted in the encounter can hardly be denied. Their comparative ignorance of mathematics occasionally led them to misunderstand the arguments of their opponents, and hence to raise objections which damaged their cause instead of furthering it. The result was that opposition from the purely philosophical side was practically silenced. On the other side, while the doctrine has become much more widely known to and studied by mathematicians in general, it has not among these met with universal acceptance: there are some who regard it as having no real significance, there are others who remain merely sceptical.

It should be added that there are points upon which the protagonists of the metageometrical theory differ from one another, and that these differences are clearly relevant to the philosophy of the doctrine. Its real significance, if it has any, thus still remains a matter of doubt, and must so remain until these differences are composed. Thus persistence in criticism of this theory, even should the criticism turn out to be mistaken, provided it be intelligent, will be justified by its utility; for it must then at least indirectly aid in bringing about that complete concordance which is still to seek in the philosophical interpretation of the theory.

With respect to Part I, the general character of which is psychological, I gladly take this opportunity of acknowledging my indebtedness to Dr. W. McDougall, Wilde Reader in Mental Philosophy in the University of Oxford, who was kind enough to read it in manuscript and to give me the benefit of some valuable suggestions.

CONTENTS

PART I

THOUGHT AND ITS SYMBOLIC EXPRESSION

CHAPTER I

CHAPTER II

CHAPTER III

PART II

IMAGINARY QUANTITIES IN ALGEBRA AND IMAGINARY LOCI IN GEOMETRY

CHAPTER IV

CHAPTER V

CHAPTER VI

CHAPTER VII

CHAPTER XVIII

CHAPTER XIX

CHAPTER XX

PART I

THOUGHT AND ITS SYMBOLIC EXPRESSION

CHAPTER I

INTRODUCTORY

Some brief account seems desirable of the way in which this work came to be written, if only to explain the connexion, not immediately obvious, between the next two chapters of it and the remainder. Indirectly, the views expressed in it grew out of a study of metaphysical and epistemological problems; but they are not, so far as I know or intend, logically consequent upon any metaphysical or epistemological doctrine.

It is a very common opinion that metaphysical disquisition is idle because it leads to no conclusions which really satisfy the mind, because the problems discussed are never really solved. This opinion no doubt springs from the observed fact that no metaphysical system succeeds in imposing itself upon thinkers generally; that scarcely has one thinker built up a doctrine satisfactory to himself than another thinker proceeds either to demolish it or to construct another inconsistent with it. Yet, for all this, these disquisitions are not idle. No one who ponders them carefully can fail, though unsatisfied in his longing for finality, to find his mental horizon enlarged as a result of the study, to find himself in possession of analogies and distinctions of thought which had previously eluded his attention.

But along with much which thus becomes intelligible there is much also which remains obscure, much which seems to consist of mere word-play, empty of real significance. And this opinion is grounded afresh and fortified in the process of critical comparison, which not infrequently yields evidence that the differences, expressed or implied, between the adherents of one doctrine and those of another, sometimes go beyond disagreement as to premises, or as to conclusions drawn from premises held in common, or as to the meanings which should be assigned to certain words or phrases (disputes about words); it becomes clear that occasionally these differences consist simply in an utter inability on the part of one disputant to assign, to propositions advanced by another, *any definite* meaning whatever.

Now when a proposition urged by one party to a discussion does not evoke in the mind of the other any definite meaning at all, is not intelligible to him, there is properly speaking no ground for agreement or disagreement; and, if the proposition is one not susceptible of any further definition than has already been given of it (remembering that explanation of the meaning of words by means of other words whose meanings are known must have its limit in the knowledge of the meaning of some words not defined), further discussion is useless. It is not that the disputants must agree to differ, but that they can neither agree nor differ.

But we may well ask how such a state of things is possible if we admit as true the presupposition which we all make, consciously or unconsciously, in the use of language, and upon which is founded the possibility of intercommunication: that there is community of thought, likeness of purely intellectual organization, between the intercommunicants; that what in experience is common to us all gives rise in us all to a common stock of abstractions, conceptions, general relations: the common basis of all intellectual construction? [1] That this presupposition may not be true is, indeed, possible, but to question it is to cut the ground away upon which rests effective intercommunication. It is a theoretical possibility which is without influence upon belief.

Rejecting this theoretical possibility of a radical difference of intellectual organization as accounting for apparently insurmountable misunderstandings, one is naturally led to inquire whether these may not arise from some hidden differences of view as to the nature of the relations between thought and language, if not more probably still from a common vagueness of apprehension of the manner in which these relations come into play. The next two chapters of this work were suggested by this question. They contain an endeavour to arrive at a fairly clear understanding of the mental attitude and processes involved in the use of language, of symbolism in general, both as a means of intercommunication and as an instrument of reasoning.

[1] I do not say 'identity' of intellectual organization, but a similarity or likeness in parallel with that of biological organization. An unintelligibility of the above kind, if not in some way bound up with the medium of intercommunication, would imply some radical difference of intellectual organization.

In the course of that inquiry my attention was forcibly arrested by the fact that the immense variety of subjects upon which people reason distribute themselves in a general way between two extremes, at one of which the process of reasoning is conducted exclusively by means of representative or typical imagery (not necessarily visual), symbols being either absent or appearing merely as habitual but useless adjuncts, while at the other extreme representative or typical images are either absent or, if present, are useless, and the process is wholly dependent upon the use of symbols.[1] The distinction does not by any means exactly correspond to that usually understood in the contrast implied by the terms abstract and concrete as applied to subjects of thought. Moreover, it appeared to me (for reasons which I cannot even briefly indicate here) that the more dependent we are on symbols for the conduct of any process of reasoning, the greater is the risk we run of illusions of judgement as to the real import of the process, of lapsing into a mystical attitude of mind with regard to it. Such an attitude of mind is no doubt quite as much and quite as often the result of indefiniteness or instability of the conceptions or ideas which form the groundwork of the subject discussed, as of the circumstance that the subject is one which demands the use of symbols for the conduct of the ratiocinative process. I am very far from saying that all mystical attitude is traceable to ratiocination by symbols, but that the necessity of ratiocination by symbols does carry in its train a tendency to mysticism from which even the keenest intellects cannot always free themselves.[2]

[1] Usually, but not necessarily, words. The term 'symbol', it need hardly be said, is employed here in a restricted and technical sense; not, e.g., in the sense of Goblet d'Alviella's *Migration of Symbols*. The symbols of that work are what would here be described as conventionalized representative images, a middle form between the purely representative and the purely symbolic.

[2] The use to which I propose to put the term 'mysticism' is, I must admit, somewhat arbitrary; but I do not know of any other name which could be less inappropriately employed to denote the kind of illusion I have in mind. What this is will become plain enough from the examples of it given in the body of the work, but even at this stage I may give some indication of the general sense in which I use the term. Some thinkers have spoken of, or alluded to, a certain 'reaction of language upon thought' which results in error or confusion, but, so far as I know, without giving any explanation of the nature of this process or any indication of the conditions under which it takes place. It is to this kind of illusion or, to

Now if we except pure or synthetic geometry there is no subject in which the conduct of the ratiocinative process is more dependent upon the employment of symbols than pure mathematics, and in which therefore, according to my view, we might more confidently expect to find evidence of the mystical tendency. On the other hand, there is no subject in which the ideas reasoned about are more definite and stable; and, certainly, mysticism, illusion as to the real import of a process of thought, seems to be utterly incompatible with definiteness and stability of the conceptions which underlie it. Merely to harbour the suspicion that mathematicians may not have developed their science in entire freedom from this tendency seems thus to betray an almost irrational leaning to scepticism. Yet, dismiss all *a priori* theoretical considerations; there are not wanting others of a practical nature which make such a suspicion at least plausible. It has long been a common-place of observation, that many men of great intellectual ability, capable in general of handling abstract subjects of thought with uncommon ease, are nevertheless apparently quite unable to learn mathematics. There is something in the subject, or in the manner of teaching it, which revolts them. I am reminded of a friend who, having taken up and dropped the study of pure mathematics, explained to me that he had abandoned the subject because, as he expressed it (with an obviously intentional touch of humour), he found that it required a kind of low cunning which he was destitute of. Expressed seriously, without whimsical exaggeration, there is in the orthodox exposition of mathematical symbolism much which, to many people, seems to be mere sophistry, paradox, and word-play. They know, indeed, that it cannot in fact be so, since contact with the practical soon makes an end of all conclusions founded upon sophistry, paradox, and word-play; and the conclusions of mathematics are in general of eminent practical application. They conclude therefore that they have essentially 'unmathematical' minds. This really means no more than that they are unable to learn mathematics, and leaves us quite

be more exact, to the mental attitude which makes it possible, that I refer when I speak of mysticism, the mystical tendency or bias. I need scarcely add that my choice of the term implies no sort of indirect judgement respecting the nature or value of the religious or spiritual experiences which have been called mystical.

in the dark as to why people with logical heads should suppose themselves incompetent to reason logically about the very few, definite, and stable concepts which are the subject-matter of the science.

With the object of reasoning commodiously and swiftly about these concepts, mathematicians have invented and gradually perfected a special symbolic system, a special language. But the construction of this language is itself a work of logic in a far more exact sense than is the case with ordinary language. The study of such a language, one would suppose, ought not to present insurmountable difficulties to minds otherwise trained to think logically. There must be something in the orthodox exposition which, if it cannot be called downright sophistry and paradox, is at least inimical to plain straightforward thinking; and the modest avowal of incompetence on the part of those who are unable to follow it is at once a tribute and a sacrifice to intellectual integrity.

In considerations such as these, no less than in those which suggested themselves to me in the course of protracted reflection on the nature of language and its functions in the process of thought, originated the supposition that even in pure mathematics, conspicuously *the* domain of definite and stable conceptions, careful investigation might be found to reveal evidence of an infiltration of the mystical. That this is in fact the case is what—after the preliminary discussion about language and symbolic expression in general—I have endeavoured to show; and I believe that this fact, if so it be, in some measure accounts for the almost insuperable difficulty which many strenuous minds find in mastering even the first principles of the subject, as well as for the purely intellectual repugnance with which they regard it.

The reader who is pressed for time but yet thinks the main subject might interest him, would perhaps do well to omit Chapters II and III as not being absolutely essential to the understanding of what comes after them. But I hesitate to recommend him to follow this course, because in doing so it is not unlikely that certain objections or difficulties may suggest themselves to him and unfavourably affect his judgement of the views expressed: objections or difficulties which probably would not suggest themselves at all after a careful perusal of these preliminary chapters.

CHAPTER II

THOUGHT AND LANGUAGE

Presupposition involved in conscious expressive action.—Thought and Meaning.—Contrast of mental attitude in the origination and in the acquisition of Language.—Beliefs suggested in the process of Language-acquisition, and more or less modified in the course of Experience.—Meaning and Definition.—Opposed views of Definition (J. S. Mill and Taine): Due to not clearly distinguishing between its nature and its function. Max Müller's doctrine of the Identity of Thought and Language.

UNDERLYING and prompting my action in writing the words now actually flowing from my pen, there is a belief or presupposition that, upon being read, they will evoke, in the reader's mind, thoughts similar to those which, in mine, seek this mode of expression. These thoughts, the reader's and mine, more or less similar as I suppose them to be, dissimilar as possibly they may be, are meanings of the written words.

This is to affirm, generally, that a meaning is a thought in an individual mind; is not a something having an existence, somewhere and somewhen, independent of any mind. But unless we agree, which I do not, to make synonyms of the words 'thought' and 'meaning', this is not to affirm the converse, that a thought or process of thinking is necessarily a meaning. And if the question arises: What differentiates a meaning from a thought? we may answer generally that a thought is a meaning just in so far as it is associated in a mind with a word or words, with marks or signs of some kind. The fact of association is what constitutes the associated things respectively meaning and symbol.

The fact itself that I thus draw attention to this presupposition or belief is evidence to others that it is at the present moment definite and explicit in my mind. This, however, is not generally the case. We commonly address one another, whether in speech or writing, without definite consciousness of the presupposition; certainly, at least, without its being explicitly present, silently embodied or formulated in words. Given the desire to communicate our thoughts, however, the act of speaking or writing, as a means to that end, would be inexplicable without the presence of the presupposition in the form of an habitual and implicit expectation, born of the long-continued experience of the actions (including speech) of those whom we address, or who are addressed by others. The habitual commonly fails to command attention, and by contrast with that which does command it, sinks into the subconscious. This mental condition has its parallel in ordinary perception. I perceive a tree; I am said to perceive that it is solid, resistant; my visual perception of solid objects is, apart from the visual element, a subconscious expectation that what is thus seen would prove to be resistant. There is no question here of deliberate inference that this would be the case; there may have been, probably there was, more or less definitely, conscious expectation when I was learning to perceive. Although we may attend to, or reflect upon, a habit of this kind, the habit itself is, in its very nature, undeliberate.

Whatever view we may take of the origin of language; whether we incline to believe in the onomatopoetic, interjectional, 'synergastic' or other original character of words, we cannot fail to recognize that language is now, and has been throughout the historic period of man's development, essentially conventional. Or, lest we risk by this statement to seem to ignore what philology has discovered for us as to the growth, decay, and transmutation of words in accordance with observed laws, it may be safer to assert that it is the *association* between word and thought which is essentially conventional; the result, not of any natural affinity between certain thoughts and certain articulate sounds, but of what may by metaphor be called repeated mechanical juxtaposition. No one, I believe, would venture to assert that there is any more obvious or natural connexion between any one than any other of the words 'dog', 'chien', 'Hund', and our conception of that animal. Nor can we discover in the constitution and use of language any reason whatever for supposing that any other articulate sound would not have served equally well with any of those in actual use as a name for the 'friend of man'. And it would in no wise modify this view if to-morrow any one should succeed in showing that these three names are indisputably traceable to some articulate sound originally mimetic in character. For it is not argued that language was conventional in its origination, but that it has become so in the course of its development.

If any one should assert that, for an infant, the primary association of articulate sounds is rather with the person who articulates than with anything else; in other words, that the infant first recognizes them merely as a proprium of persons, it would be hard to find any good reason to controvert the assertion. Point to any object and name it; even if the infant sees the object, the primary association will be rather between your gesture and your articulation than between either of these and the object—unless, indeed, you credit the infant with an *a priori* knowledge of the meaning involved in pointing, of the intention which underlies that act. You might then also just as well endow him with an *a priori* appreciation of the natural fitness of the names he hears for the things he sees and handles. Throw a stick or a stone for an untrained dog to fetch. If the object thrown catches the dog's eye he may possibly run after it; but if not, you may point in the direction of the object with no other result than the dog's wagging his tail, barking, and capering; and the more emphatic your gesture, the more vigorously he will wag, bark, and caper. The dog has not learnt your intention in pointing, does not associate that particular gesture with the object to be fetched. The man, or the man-like ancestor of man, who first pointed to an object with the deliberate purpose of directing to it the attention of his fellows, might well be looked upon as the founder of the intercommunication of thought upon a conscious and quasi-artificial basis. It is true that this gesture, like the facial movements, the ejaculations, and other gestures which accompany emotion, may have been instinctive; may, like them, have originally served the unpurposive intercommunication of rudimentary thought. But the critical and decisive step towards symbolism of any kind (oral, gesticulatory, mimetic) is taken when deliberate or purposive use is made of these physical adjuncts to mental states. The way then lies open to ingenuity for a progressive and gradual fashioning of the raw material of expression, viz. personal action. And this handiwork of man is apparently subject to no control more definite than that of racial characteristics, varying conditions of life and varying nature of the environment. Thus in the lapse of ages, as the original and instinctive is replaced in diverging directions by the artificial and conventional, we find the denizen of the West manually expressing respect by the conventional act of removing his hat, while the Oriental expresses this by the equally conventional act of removing his shoes—a difference which is no more and no less striking and significant than that of the articulate sounds (the verbal action) in which they respectively embody the sentiment which moves them both.

While the origination and elaboration of language demand purpose and ingenuity, the learning of language is by comparison mechanical; it is a process of association which, in its earliest stage, must be almost entirely passive, taking place independently of definite purpose in the child's mind. The articulate sounds which the infant hears are at first merely natural phenomena, probably not even at once connected with the persons who utter them. A long and laborious process must be gone through before these articulate sounds become for the child what they are for us, words or symbols. The first stage in this process is probably their recognition merely as a proprium of persons. We know, as a matter of fact, that in course of time they become associated in the child's mind with the things, events, actions, qualities, &c., whose names we say they are. I only wish to suggest here that this process of association is attended, on the side of the articulate sounds, with a process of abstraction similar to that which goes on with respect to the phenomena with which they become associated. What we call a word, say the word 'dog', is no more and no less the articulate sound which I utter than it is the articulate sound which you utter; it is an 'identity in difference' of articulation and intonation which the child recognizes. An articulate sound has, of course, no existence apart from these peculiarities of personal utterance, but as the purport and intent of the process of association becomes understood, these peculiarities at the same time become ignored as non-essentials, and the word, thus abstracted from what is a proprium of persons, tends, like all other abstractions, to assume an independent existence.

It may be remarked in passing that if the import and purpose of names are disclosed only if there is the ability to make abstraction of (1) the peculiarities of personal utterance, (2) the differences which constitute the several individualities of the similar objects, events, &c., to which the same name is applied, in proportion, that is, to the formation of abstract ideas and simple concepts—then it is difficult for us to accept as adequate to the facts the oft-repeated assertion that the possession of language is a pre-requisite to the formation of abstract ideas and simple concepts. It would seem rather that one of the conditions of the apprehension of language, not to speak of its origination, is the capacity to form these ideas and concepts.

Many of the properties of objects and of the results of action the child discovers unaided, but no amount of observation or assiduity of experiment can disclose their names: this is a secret known only to the 'grown-ups'. How the grown-ups have come by this knowledge is a question which probably no child ever asks himself; I have never read or heard of a child endeavouring to find out from any one how things come by their names. For him (as for philosophy in its infancy) the names of things, though they can only be revealed, not discovered, are 'natural' signs. Evidently the modern proximity of various races, civilizations, and languages, makes it extremely difficult for such a view to maintain its hold upon the individual's mind as his experience widens; yet the customary always seems natural, not conventional; so powerful is the effect of this customary association between name and thing, that many individuals never fully realize the radical conventionality of language, and in the extreme case the vacant grin of the yokel attests his sense of the absurd in the foreign name of any familiar object.

When, after long mental floundering, there is at last formed in the child's mind the general notion of a sign or mark, and the clear recognition of articulate sounds as phenomena invested with that character, the process of learning language gradually merges from the relatively slow and mechanical to the rapid and intelligent. This marked change in readiness of apprehension is often very striking and has not escaped the notice of those who observe closely the mental development of children. There is almost certainly involved in this change of mental attitude from the merely receptive to the actively inquiring the formation of an implicit expectation (one of those beliefs, inductions from experience, which remain unquestioned, unreflected upon, so long as no case occurs of unfulfilled expectation), that everything has a name. Children, arrived at this stage, frequently ask what a thing is called; I do not think they ever ask whether a thing has a name or not.

A somewhat similar change is noticeable at a later stage of intellectual development, when the child first distinctly forms the notion of cause, or perhaps of explicability, and the incessant and insistent Why? distracts its elders, while it may flatter them as evidence of so genuine a belief in their omniscience. Further, it is to be remarked—and the point is of importance in relation to this inquiry—that just as children in the *Why?* phase expect that a reason or cause can be assigned for everything, so, at the earlier stage we have been considering, there is formed an implicit belief that the fact or use of a name involves the existence of something (object, person, event, &c.) corresponding to it: a postulate which becomes quite spontaneous though to some extent corrected by experience, and which, though it has rarely been the origin, has often been the confirmation, of the marvellous, the superstitious, the mythological. Max Müller, for whom mythology was merely a disease of language, but whose views on that subject did not commend themselves to philologists and students of mythology, in effect saw in this postulate the origin, instead of the support, of mythical belief.

Subsequently, when the conception of mind as distinct from body, of persons as not only actors but thinkers, breaks in upon the child, and words are recognized as expressions of thought as well as signs of things, this postulate receives an extension. Words denoting abstractions, and collocations of words expressive of the combinations of abstract thinking, which the child hears but does not understand, become for him unquestioned evidence of the existence of ideas and ideal constructions corresponding to them; and whatever these expressions evoke in his own mind, however vague and indefinite, he is often content to take as their meaning. It is one of the commonest experiences with children at a certain stage of development to find them using forms of speech, phrases, expressions which they have picked up from the conversation of their elders and which they bring forth again more by chance verbal association than as expressing definite ideal constructions. This habit or tendency in the child, if it survives, and so far as it survives, in the adult, must clearly be a potential source of illusion.

Stability in the relation of meaning to symbol is one of our chief intellectual needs, whether we consider the individual or the group of intercommunicants—so great a need, indeed, that the common tendency is silently to accept this stability as an actual fact rather than as a desideratum. For the ordinary requirements and practical concerns of life the actual stability of this relation is accepted as sufficient. In current oral communication the formal language common to the group of intercommunicants is aided in its task by informal expression and the familiarity with personal idiosyncracies; we frequently understand almost as much by the eye as by the ear. But from the moment we pass from the ordinarily colloquial to the severely accurate, academic, or scientific treatment of a subject, the necessity of an exact and adequate terminology at once makes itself felt; and, with exactness in terminology we become especially concerned with formal definition.

Definition is, itself, a word of which lax use is often made. The definition of a word, for instance, is often confounded with its meaning: a confusion which is commonly immaterial for our ordinary purposes, since the object of defining is to convey meaning, but which is damaging to clearness of apprehension in the subject which now concerns us. It will serve to render more definite the standpoint from which I regard the relation of language to thought if I here make a few remarks on the subject of formal definition, and on the distinction which it seems well to observe between definition and meaning.

It was long ago observed by Reid, in his *Essay on the Intellectual Powers*, that, 'A definition is nothing else but an explication of the meaning of a word, by words whose meaning is already known. Hence it is evident that every word cannot be defined; for the definition must consist of words; and there could be no definition, if there were no words previously understood without definition.'

Formally, of course, every word is susceptible of definition, as we see in the dictionaries; but although we may learn the meanings of many words through definition, we have each of us *ab initio* to learn from others the meanings of some words by a process of guesswork which results in right association.

A dictionary of a language, which is very often regarded as a collection of the words of the language and of their meanings, is, more strictly, a collection of these words and of their definitions. The point of the distinction lies, for me, in this: that the meanings of the words are by metaphor only in the dictionary, while in reality they were in the mind of the lexicographer, and may be evoked in the mind of any one who consults his book. This view leads one somewhat further than is at first sight quite apparent. The defined and the defining words alone are in the dictionary. But the symbols which, in the mind, are associated with and are said to express our thoughts, are clearly not in the dictionary, since they are in the mind. These symbols are words. Hence it seems logically to follow that the printed characters in the book are words only by metaphor; that words, like the thoughts of which they are symbols, exist in minds only. This conclusion is, of course, totally opposed to the customary view; but after all it is, itself, merely a matter of good or bad definition, of a consistent or inconsistent use of language from a philosophical or scientific standpoint. It is clear that if we apply the same name to something in a mind and to something in a book we apply the name metaphorically either in the one case or in the other. Which is the metaphorical use? I do not think it much matters what answer we give to the question; but so far as it can be said that a right answer to it there is, I think it will be found that, from a purely philosophical point of view, it is more consistent to consider bodies or material objects (such as printed, written, or carved characters), and personal actions (such as speech and the gestures of the deaf and dumb), as being words only by metaphor.[1]

Returning to the distinction between meaning and definition, we may put it thus: Although we ask for, and expect to get, the meaning of a word, all that can possibly be given to us, in words, is its definition; it then rests with us whether or not we possess its meaning. All that, in this respect, we can either afford to, or receive from, others is definition: meaning cannot be directly communicated.[2] Thus, although definition implies, involves, and by metaphor is often said to *be* meaning, a definition should no more be confounded with the meaning of the word defined than should the several words of which it consists be confounded with their respective meanings. If we hold fast to this distinction we are in a position much more readily to understand the opposition of view in which eminent thinkers have

[1] See pp. 18, 19.

[2] This last assertion will have to be modified when, if ever, 'thought-reading' is established as a fact.

found themselves on the subject of the real nature of definition. Take, for instance, Mill's opinion as expressed in his *System of Logic*, and the view which Taine affirms in the short but valuable criticism of Mill contained in the *Histoire de la Littérature anglaise*.

Mill, contesting the then prevalent notions, writes:

'The definition, they say, unfolds the nature of the thing: but no definition can unfold its whole nature, and every proposition in which any quality whatever is predicated of the thing, unfolds some part of its nature. The true state of the case we take to be this. All definitions are of names, and of names only; but, in some definitions, it is clearly apparent, that nothing is intended except to explain the meaning of the word; while in others, besides explaining the meaning of the word, it is intended to be implied that there exists a thing, corresponding to the word.'

Taine replies to this as follows:

'D'abord la définition. Il n'y a pas, dit Mill, de définition de chose, et quand on me définit la sphère le solide engendré par la révolution d'un demi-cercle autour de son diamètre, on ne me définit qu'un nom. Sans doute on vous apprend par là le sens d'un nom, mais on vous apprend encore bien autre chose. On vous annonce que toutes les propriétés de toute sphère dérive de cette formule génératrice. On réduit une donnée infiniment complexe à deux éléments. On transforme la donnée sensible en données abstraites; on exprime l'essence de la sphère, c'est-à-dire la cause intérieure et primordiale de toutes ses propriétés. Voilà la nature de toute vraie définition; elle ne se contente pas d'expliquer un nom, elle n'est pas un simple signalement. . . . La définition est la proposition qui marque dans un objet la qualité d'où derivent les autres, et qui ne derive pas d'une autre qualité. Ce n'est point là une proposition verbale, car elle vous enseigne la qualité d'une chose.'

There is, of course, much that is true in both these views, notwithstanding their opposition. But would there not have been much less if both Mill and Taine had clearly recognized the distinction upon which we have been insisting? Thus, we must agree with Mill that all definitions are of names, and of names only, provided our intention is to mark the distinction between knowledge about the use of words and knowledge about 'things', and not to assert that knowledge about the use of words is a practical possibility apart from knowledge of what the words signify, that is, meanings. If Mill had confined himself to pointing out that the object of definition is to evoke a meaning which the defined term has so far not evoked, and that we ought not to confound definition with its object, a means with an end, there would have been nothing to say. But he goes on to confuse the issue by irrelevant considerations regarding the existence of the things defined. The existence or non-existence of a thing defined is not what makes the difference between some definitions and others. From Mill's own point of view, that definitions are of names only, the difference between one definition and another or between the same definition in different minds, is rather this: that in one case the definition may evoke meaning, in the other not. This, however, does not affect the nature of definition itself. Thus, I may be given the definition of 'allochiral', as that similarity which the right hand has to the left: similar, that is, with reversed parts. Now, I may have been quite familiar with this peculiar kind of similarity of shape without knowing that any name had been invented for it. In that case I shall merely acquire the knowledge of a name. But if this definition evokes in my mind the awareness of a similarity previously unnoticed, I shall have acquired, besides the knowledge of a name, a knowledge of 'things'. In both these cases the definition evokes meaning; but it might perfectly well be that the defined word had reference to some branch of knowledge with which I was totally unacquainted, that the definition given of it was itself highly technical and evoked no more definite meaning in my mind than did the term defined by it. I should then be in possession of a barren piece of information, viz. that the defining words may be substituted for the word defined. In none of these cases is the essential character of definition as such at all modified; the one common point is that a name is defined in all of them, whatever else may attend or follow upon the definition.

Turning to Taine's criticism, it appears that he misses the point somewhat obscurely put by Mill. He distinguishes between knowledge about words and knowledge about things, but he fails to tell us how we are, verbally, to mark the distinction. If we are to use the same term, viz. 'definition', to denote both kinds of knowledge, we court confusion, at least in philosophical discussion, where it is important to maintain the distinction. Mill's plea is, in substance, that we should avoid this confusion by restricting the reference of definition to knowledge about words. Taine, on the other hand, had his mind fixed on the import or value of definition, and protested against what he supposed was a subordination of the office of definition. He is concerned about *la vraie définition,* i.e. that which gives the essence of the thing defined, and his remarks are valuable, but they are directed to a different point from that which concerned Mill.

What gives an air of unreality to this particular discussion, and perhaps also to Mill's contention, is that the nature of definition is, for all ordinary purposes, of so very little importance as compared with its function. Evocation of meaning, and not a bare verbal knowledge, is what interests us and engages attention. If we agree with Mill that in strictness definitions are of names only, it must also be admitted that this bare knowledge we hardly ever seek. The definition of a name is of no use to me save as a means of discovering that which, in the definer's mind, evokes the name and the definition he gives me of it.

In the interest of general intelligibility the individual is not warranted in arbitrarily assigning meaning to words by definition; but, in the same interest, it should be our common and concerted endeavour, by that means, to dispel the obscurity which arises from lack of decision in the use of words, that is, from instability of association between word and thought. Although it is custom —the external evidence of the common will—which finally sanctions the use of any word in a particular sense, or in different senses, yet it is reasoning, good or bad, and frequently acting through the paramount authority of individuals, which initiates custom. The ordered development of articulate sounds, indeed—phonesis as distinguished from symbolism—is undoubtedly the result of physical forces which exhibit, in this result, certain more or less definite laws of action; it is therefore properly regarded as lying outside the sphere of ratiocinative action. But symbolism is essentially the work of intelligence. The use which we make of artificial signs is throughout largely directed by our perception of resemblance and analogy, and our discernment of the value of metaphor as a means of expression.

When from the ordinary, not the psychological, standpoint we ask the question, What is a word? we come back to the question discused a few pages back. Should words be regarded as objective or as subjective?—in one sense a question as to the right, i.e. the most convenient, use of the term 'words' itself. I said in effect that from a philosophical standpoint words should be regarded as mental entities, as an integral part or domain of thought. We might put it, that just as we are careful to distinguish percepts and images from the objective phenomena which are their physical correlates (and the fact of distinction is common ground, whatever metaphysical theory we may adopt as explanatory of it), so we ought to distinguish between words and their physical correlates. But, besides thus begging the question, we also thus arbitrarily and in defiance of custom limit the use of the term. Custom not only sanctions but enjoins the application of the term to those physical objects which are the characters printed on this page, and to those physical actions in which speech consists; and to custom it is necessary, in language above all, to conform. But we must then admit that we use the term in a metaphorical as well as in a literal sense. Which of the two we should look upon as the literal and which as the metaphorical is of less importance than not to lose sight of the distinction. As we shall see later on, failure to take into consideration either the objective, or again the subjective, aspect of words leads to extreme and irreconcilable theories of the relation between thought and language.[1]

Nevertheless, for philosophical purposes, it seems better to consider the subjective aspect as the literal meaning, for the reason that the literal meaning of any term should be the least ambiguous. Now from the moment we observe the distinction between the subjective and objective aspects of words, it becomes at once evident that the term does not always lend itself to clear interpretation as the name of an objective entity. How is this entity to be unambiguous and simply identified in speech? Is it the action of the articulatory organs, or the sequence of aerial undulations, or the particular vibrations of the tympanum, or any other of the physiological processes concerned; or is it the entire series of these actions? In its subjective aspect, on the other hand, a word is a perfectly definite entity, psychologically analysable, indeed, into constitutive elements, but none the less a definite entity.

Before I bring this chapter of more or less general considera-

[1] Cf. chap. iii, pp. 48 et seq.

tions to a close, I must very briefly criticize a well-known theory of the connexion between language and thought which is wholly incompatible with the views I hold. I mean the doctrine that thought is impossible without language: that thought and language are identical. This view, as is well known to all those interested in the philosophy of language, was consistently maintained by Max Müller, and defended by him to the end against all objections. Had a really genuine defence of this theory seemed to me possible, I could not have attempted what in this work I have attempted: I would have felt I had no firm ground to stand on. For it is precisely, though not solely, to an insufficiently vivid awareness of the distinction between meaning and the symbols in which it is expressed that I attribute the occasional intrusion of mysticism into close logical reasoning. This being the case, I suppose I ought to say something about the doctrine in question. I might urge, indeed, that Max Müller's views have already been adequately criticized and, in my opinion, refuted; in particular by the late Professor Whitney in that able little book, *Max Müller and the Science of Language.* But the question being one which has so precise and important a relation to the subject of this work, I will indicate my own reasons for dissent, glad though I am to claim the support of a recognized authority. In order to be as brief as possible I shall consider those points only, advanced by Max Müller in defence of his theory, which are absolutely essential to its stability. The quotations are all from *The Science of Thought.*[1]

'I mean by thought the act of thinking, and by thinking I mean no more than combining. . . . "I think" means to me the same as the Latin *cogito,* namely, co-agito, "I bring together" . . . combining means separating.'

Next, as to the materials or elements of Thought, we have:

'(1) Sensations (Empfindungen),
'(2) Percepts (Vorstellungen, presentations),
'(3) Concepts (Begriffe),
'(4) Names (Namen).'

The first chapter of *The Science of Thought* is devoted to showing that these elements of thought are distinguishable by analysis, but do not exist in the mind as separate entities; and

[1] *The Science of Thought*, by F. Max Müller, 1887.

the character of the arguments used makes it clear that what we are to understand by this is that these several elements of thought necessarily involve one another, or exist in the mind only as interdependent. It must be remarked, however, that in this first chapter no sort of proof is offered that names are involved necessarily with the other elements of thought. This is, or seems to be, assumed. Proofs or reasons are tendered subsequently. We seem thus to start with an assumption, to be subsequently validated, that thought and language are inseparable, in the sense that thinking without words is impossible. That this is the real nature of the assumption is clear from the following passages. After adducing the names of a number of eminent philosophers in support of his view, Max Müller proceeds :

'But we have now to ask the question, which to my mind is most perplexing, How was it possible that not one of these philosophers, not even those who fully recognized the inseparableness of language and thought, should have seen that this discovery of the true relation of language and thought, or what may truly be called this revelation of the oneness of thought and language, means a complete revolution in philosophy ? '

He then adds, a little further on :

'I have freely and fully admitted that thought may exist without words, because other signs may take the place of words . . .' [1]

There is of course no real contradiction here. The two passages, taken together, serve to show in what sense we are to understand the 'inseparableness' : thought has no existence apart from signs. The inseparableness is not merely an inseparableness which is the result of artificial and long-continued association between sign and thought, but an inseparableness *a priori*, an essential identity or oneness.

But, in fact, this proposition of the 'oneness of thought and language' never gets beyond the stage of assertion. We may agree that sensation, perception, and conception are all necessarily involved together, are never really isolated, in the process of human thought. But on what grounds are we to accept as true the allegation that names are necessarily involved with these other elements of thought ? I have sought them in vain in

[1] *Op. cit.*, pp. 50, 51.

The Science of Thought. We never seem to advance beyond mere assertion, or a disguised begging of the question. Here, in the following abridged excerpt, we have an instance:

'We now come to the . . . most fiercely contested question, namely, whether concepts can exist without words. . . . We mean by language what the Greeks called Logos, word and meaning in one, or rather something of which word and meaning are only, as it were, the two sides.' [1]

But, evidently, if the relation between thought and language is a subject which requires investigation, we are not justified in settling the question off-hand by means of a definition. We thus in effect close the door upon real investigation.

To those who, judging from analogies of individual action, conclude that speechless infants and some of the higher animals think, however rudimentarily, Max Müller addresses a warning against the 'dangers of Menagerie and Nursery Psychology'—a telling phrase from the controversialist's point of view. But we must refuse to take it for more than it is really worth, that is, as a caution against rashness in analogy, not as a prohibition of its use. Upon this point Max Müller seems not to have been very clear, if we are to judge from the following statement:

'It has often been said that animals have sensations and percepts, but that one ought not to ascribe to them the possession of concepts. Of the conventional animal of the philosopher this may be quite true. We have a right to conclude from analogy that it is so, provided only that we . . . admit that we do not in the least know how animals philosophize, and how an ox recognizes his stable door.' [2]

I suppose this means that we may conclude from analogy, provided we admit that the conclusion is without value; in other words, that analogy is here valueless. Yet, after all, it is only upon the same *kind* of analogy that rests our knowledge of one another's thoughts. If I do not *in the least* know how an ox recognizes his stable door, I seek in vain for any valid reason to suppose that I know how my neighbour recognizes his house door.[1]

The question, once much discussed, whether names are names of concepts or names of things, is revived by Max Müller in connexion with the general subject. He agrees with J. S. Mill in

[1] *Op. cit.*, p. 29. [2] Ibid., p. 27.

making names signs of concepts, not of things. So far as I know, neither side in this discussion appears ever to have accorded the least consideration to what is, after all, the essential part of the matter, viz. the purpose and intention of the namer; and this, as disclosed by custom, is not doubtful. Names are used at will as signs both of concepts and of things, and the reason of this obviously lies in the interdependence of percepts and concepts. Max Müller's argument runs as follows:

'Words are, in their origin, the signs of concepts and not of things. . . . If we use a general name, if we say Dog, do we mean the thing or the concept of it? Is there anything corresponding to Dog? Is not Dog, like every other name, the name of a thing which cannot possibly exist? Who ever saw a dog? We may see a spaniel, or a greyhound, or a dachshund, we may see a black or a white or a brown dog, but a dog no human eye has ever seen. Therefore, when we say dog, we can only mean our concept of a dog, that is our concept of many or all dogs, and it is the name of that concept which we use to denote any single dog.'[1]

But if the *raison d'être* of a name is to denote anything, and if the name which denotes the concept is also used 'to denote any single dog', why is it not also the name, or rather *a* name, of that single dog? It is strange, too, that Max Müller should not have seen that he would not be allowed to stop short at spaniels, greyhounds, &c. If no one ever saw a 'dog', then certainly no one ever saw a 'spaniel', nor even 'a black spaniel', nor yet 'a large black spaniel', though any one who pleases may see my large black spaniel 'Gypsy'. And in answer to the question: Is there anything corresponding to Dog?—I see no reason why we should not answer, (1) that there are a number of things of each of which 'dog' is *a* name, and (2) that there is our awareness of resemblance between these things, i.e. our concept, of which 'dog' is *the* name.

A name, then, is a sign which I associate with every one or any one of a number of percepts, actual or reminiscent, simple or compound, in accord with the particular concept which involves them, and which I also associate with the name. It is true that upon the temporary purpose of my thinking, upon the particular kind of resemblance which engages my attention depends the particular name which I associate with the percept; and this

[1] *Op. cit.*, p. 78.

particular name (be it 'dog', 'quadruped,' 'animal,' &c.) being *uniquely* associated with the concept denoted by it, is *the* name of the concept, while it is only one of the names of the percept or of the thing perceived. To this extent it is possible to agree with Mill and Max Müller. In short, a general name, being the name of a concept, is general just because it is applicable in turn to each of the particulars which involve it and are involved in it. Bishop Berkeley, who supposed that the admission of the existence of abstract ideas was an obstacle to the establishment of his theory of idealism, used precisely this argument in the mistaken endeavour to show that abstract ideas, or concepts, are figments. A general name, according to him, is *nothing but* a name successively applied to particulars.

The theory of the essential identity of thought and language, if true, would deal a staggering blow to the Darwinian doctrine of a community of descent between man and monkey. Conversely, the latter doctrine is so obviously hostile to Max Müller's theory, that he endeavours to controvert Darwin on this particular point. After affirming his general agreement with Darwin in principle, he proceeds:

'I differ from him, however, when we come to the question of the descent of man from some unknown ancestor, because I look upon language as a property of man of which no trace, whether actual or potential, has ever been found in any other animal. I therefore contend that Darwinians, if true to the principles enunciated by Darwin himself, ought to accept the conclusion that man cannot be descended from any other animal, provided always that I can establish my premiss that language is really a proprium of man and of man only.

'If we speak simply of the materials, not of the elements, of language—and the distinction between these two words is but too often overlooked—then, no doubt, we may say that the phonetic materials of the cries of animals and the languages of man are the same. . . . But even after having traced back some at least of the material elements of language to interjections and imitations, people ought not to imagine for one moment that they have thus accounted for the real elements of language. We may account for the materials of many things, without thereby accounting for the functions which they perform. If we were to take a number of flints, more or less carefully chipped and shaped and sharpened, and were to say that these flints are just the same as other flints formed by thousands in fields and quarries, this would be about as true as to assert that the materials of language are the same as the cries of animals or, it may be, the sounds of bells. And if I were to say that apes had been seen to use flints (*The Pavians in Eastern Africa*; see Caspari, *Urgeschichte*, vol. i, p. 244), that they could not have helped discovering that sharp-edged flints were the most effective, and would therefore have either made a selection of them or tried to imitate them, that is to say, to give to raw flints a sharp edge—that therefore the presence of highly finished flints does in no way prove the presence of man—what would the antiquaries say to such heresies? And yet to say that no traces of human workmanship can be discovered in these flints, that they in no wise prove the early existence of man, or that there is no insuperable objection to the belief that these flints were made by apes, cannot sound half so incongruous to them as to be told that the first grammatical edge might have been imparted to our words by some lower animals, or that, the materials of language being given, everything else, from the neighing of a horse to the lyric poetry of Goethe, was a mere question of development.'[1]

The argument is ingenious; but its ingenuity lies in the avoidance of the real point at issue. Remark, first, that in this passage Max Müller was not careful to observe the distinction, upon which he himself insisted, between the materials and the elements of language. If it is erroneous to say that the chipped and shaped flints are the same as the flints found by thousands in field and quarry, then the parallel error would lie in an assertion that the *elements* of language, i.e. the fashioned or 'grammatically-edged' materials, are the same as the cries of animals. Next, let us suppose that corroborative evidence were forthcoming of the use of flints by apes. I can see no reason why the antiquaries should allow themselves to be disturbed by this newly established fact, nor why they should not listen with perfect composure to the supposition that apes might select flints and even try to imitate those selected. Nor, again, admitting the supposed facts, why they should think it a heresy to say that the existence of such flints in any particular locality was no proof of man's early existence there. But they certainly would, or ought to, reject any implied conclusion that because apes could do these things, therefore an ape like prehistoric man could not have existed, or could not have descended, together with such apes, from a common ape-like ancestor.

The theory of the unity or identity of thought and language, in

[1] *Op. cit.*, pp. 175, 176.

the sense intended by Max Müller, is certainly not established by the arguments which he brought forward and urged in support of it. But of course it may be said that although he failed to make good his case, the doctrine may none the less be true. The question will force itself to the front once more in the discussion of language as an instrument of reason: the subject of the following chapter. There we shall find Max Müller defending his theory against objections of an even more formidable character than those which arise from the general considerations which have occupied us from the beginning to the end of the present chapter; and I think it will have to be admitted that the arguments by which he seeks to defend his doctrine against these objections are precisely of the same nature as those by which he sought to establish it: when stripped of all irrelevances they are seen to be mere assertion and re-assertion of the doctrine.

CHAPTER III

LANGUAGE AS AN INSTRUMENT OF REASON

Sir William Hamilton on Language as an aid to Thought.—Ambiguity in the statement of the problem.—Professor Whitney on the same subject.—Analysis of the merits and defects of Whitney's view.—Verbal and mathematical symbolism.—Professor Stout's distinction between the function of words and that of substitute signs.—This distinction too trenchant: there is community as well as difference of function.—Controversy in *Nature* on the subject of 'Thought without words'.—Conclusion as to the mode of operation of symbols in aid of the process of thought.—Apparent agreement with Professor Stout's views.—Objective and subjective aspects of Language.

In Sir William Hamilton's *Lectures on Logic* there is a passage on the aid afforded by language to thought which is probably familiar to most of those who have studied his works.

'Language—says Hamilton—is the attribution of signs to our cognitions of things. But as a cognition must have been already there, before it could receive a sign, consequently that knowledge which is denoted by the formation and application of a word must have preceded the symbol which denotes it. . . . A sign is necessary to give stability to our intellectual progress—to establish each step in advance as a new starting-point for our advance to another beyond. A country may be overrun by an armed host, but it is only conquered by the establishment of fortresses. Words are the fortresses of thought. They enable us to utilize our dominion over what we have already overrun in thought; to make every intellectual conquest the basis of operations for others still beyond. Or another illustration: You have all heard of the process of tunnelling, of tunnelling through a sandbank. In this operation it is impossible to succeed unless every foot, nay, almost every inch, in our progress be secured by an arch of masonry, before we attempt the excavation of another. Now, language is to the mind precisely what the arch is to the tunnel. The power of thinking and the power of excavation are not dependent on the word in the one case, on the mason-work in the other; but without these subsidiaries, neither process could be carried on beyond its rudimentary commencement.'[1]

Mill quotes this passage somewhere in his *Examination of Sir*

[1] *Op. cit.*, vol. i, p. 138.

William Hamilton's Philosophy, and accords it the praise which it certainly merits for the aptness and felicity of its illustrative analogies. But we must not be uncritical in our admiration. We have to ask whether these comparisons really disclose analogy of action, or merely suppose it; whether they in fact make clear to us the way in which the verbal process subserves the thinking process. We understand how the establishment of fortresses subserves the conquest of a country. The mechanism by which the mason-work makes possible the process of tunnelling through sand is plain to us. But do we thereby at all grasp the mode of operation of words in the facilitation of thought? How does the naming of a cognition render the possession of it more secure and lasting? And how does the use of names for images and concepts aid, or enable us to carry on, beyond its rudimentary commencement, any process of reasoning?

At first sight one would suppose that the naming of cognitions, the attribution of symbols to images and concepts, would constitute merely an additional burden on the memory, necessary indeed for the purpose of effective intercommunication, but a hindrance rather than a help in the prosecution of a train of thought. Although self-introspection does not in general confirm this view, and is very far from doing so with most people, there are some who very definitely assert it. Mr. Francis Galton, for instance, tells us that 'It is a serious drawback to me in writing, and still more in explaining myself, that I do not so readily think in words as otherwise.'[1]

We can hardly regard the matter as one merely of personal idiosyncrasy, and those will be the less inclined to do so who have had experience of the difficulties which attend us, and of the illusions to which we are exposed, in the interpretation of evidence afforded by self-introspection. The conflicting views would tend to harmony if we could admit that the difficulty, if not the impossibility of thinking without words experienced by so many people, is no more than the difficulty or impossibility of severing an association between sign and signification which has become spontaneous through the learning of language and the constant intercommunication, by means of words, between mind and mind. To admit this, however, would be to admit that the more general interpretation of the evidence afforded by

[1] See p. 40 (Quotation from Galton).

self-introspection is erroneous. The truth of the matter is, I believe, that neither of the interpretations is wholly adequate to the facts. Admitting, as I think we must, that peculiarities of mental habit are not altogether negligible factors in seeking an answer to the question, Is language an aid in the process of reasoning ?—yet, in the main, that answer is independent of them. We ought rather, as will be seen later, to put the question thus : Does the presumed aid which symbols lend to the process of reasoning depend upon the content of that process, upon the nature of the concepts which are involved in it ?

Hamilton does not really inform us. Analogies which, however ingenious, merely throw a possible hypothesis into the form of an explanation, or which do little more than restate a problem in the guise of an answer to it, are not here of much use to us.

When we say, in general terms, that language is a necessary instrument of human thought, we must remark that this statement may express the assertion of two very different propositions. We must clearly distinguish between the proposition that language renders service to the human intellect through the dissemination and the perpetuation of knowledge acquired at first hand, and the proposition that it affords aid to the individual in the conduct of the reasoning process. Otherwise we shall be in danger of accepting illustrations and explanations of the one proposition as illustrations and explanations of the other. Moreover, the second proposition, that language affords aid to the individual in the conduct of the process of reasoning, is itself susceptible of two interpretations if we include, under the term 'language', the written as well as the oral form. Every one understands how a process of reasoning—especially when long and complicated—is facilitated by embodying it, as it develops, in external and permanent symbols. It is quite another matter to understand how the internal symbolization of such a process as it develops aids the process itself.

Having heard what an eminent metaphysician and logician had to tell us on this subject, let us now consider what a distinguished philologist can contribute to it. I take Whitney's well-known work, *Language and the Study of Language*. Under the heads 'Difference of Mental Action in Men and other Animals' and 'Aid given by Language to Thought', pp. 415–21, will be found the following remarks:

'It has often been remarked that the crow has a capacity to count, up to a certain number. If two hunters enter a hut and only one comes out, he will not be allured near the place by any bait, however tempting; the same will be the case if three enter and two come out, or if four enter and three come out—and so on, till a number is reached which is beyond his arithmetic. . . . Something very like this would be true of men, without language. Open for the briefest instant one hand with one corn in it, and then again with two, and any one who has an eye can tell the difference; so with two and three, with three and four—and so on, up to a limit which may vary with the quickness of eye and readiness of thought of the counter, results of his natural capacity or of his training, but which is surely reached, and soon. Open the hand, for instance, with twenty corns, then drop one secretly and open it again, and the surest eye that ever looked could not detect the loss. . . . But here appears the discordance between the human mind and that of the brute. The crow would never find out that the heap of twenty is greater than that of nineteen; the man does it without difficulty: he analyses or breaks up both into parts, say of four each, the numerical value of which he can immediately apprehend, as well as their number; and he at last finds a couple of parts, whereof both he and the crow would see that the one exceeds the other.

'In this power of detailed review, analysis, and comparison, now, lies, as I conceive, the first fundamental trait of superiority of man's endowment. But this is not all. This would merely amount to a great and valuable extension of the limits of immediate apprehension; whereas the crow knows well that three corns are more than two corns, man would be able also to satisfy himself, in every actual case which should arise, that twenty corns are more than nineteen corns, or a hundred corns than ninety-nine corns; and he would be able to make an intelligent choice of the larger heap where a crow might cheat himself through ignorance. So much is possible without language, nor would it alone ever lead to the possession of language. In order to this, another kind of analysis is necessary, an analysis which separates the qualities of a thing from the thing itself, and contemplates them apart. The man, in short, is able to perceive, not only that three corns are more than two corns, but that three are more than two—a thing which the bird neither does nor can do. Such a perception makes language possible—for language-making is a naming of the properties of things, and of the things themselves through these properties—and, combined with the other power which we have just noticed, it creates the possibility also of an indefinite progression in thinking and reasoning by means of language. Signs being found for the conceptions "one", "two," "three," and so on, we can proceed to build them up into any higher aggregate that we choose, following each step of combination with a sign, and with that sign associating the result of the process that made it, so as to be effectually relieved of the necessity of performing the process over again in each new case.'

We may, I think, put it thus: Images are, in general, inadequate to *represent* number-concepts. The artifice of naming or symbolizing numbers springs from a co-ordination of the concept of number in general with that of order, the essential trait in the invention of numerical names or symbols lying in their fixed serial association in memory, so that the relations of temporal or spatial order between the names or other symbols give us the relations of number between the aggregates symbolized.

A complete and convincing explanation of the difference of mental action between man and brute would evidently contain a complete explanation of the relations of thought and language. But even in the mere fragment of an explanation which Whitney gives us of this difference of mental action there is room for a good deal of scepticism. It is easy to admit that in the power of detailed comparison and analysis lies the fundamental trait of man's superiority of endowment over the brute. But what are the grounds on which rest the assertions that man separates the qualities of a thing from the thing itself and contemplates them apart, while other animals are unable to do this; that man perceives three to be more than two, while the crow neither does nor can perceive this? Take the latter of these two statements first. If it be admitted that the crow perceives three hunters to be more than two hunters, three corns to be more than two corns, three crows to be more than two crows, and so forth, is not this to admit that the crow is aware of that difference between aggregates of men, corns, crows, &c., which we say is the difference between three and two? I do not see how, with any assurance, we are to deny this knowledge to the crow. And, to make the denial more difficult, we have only to consider that the crow refuses to be drawn when *any* three hunters enter the hut and only two leave it. He knows, then, that *any* three hunters are more than *any* two hunters, that *any* three corns are more than *any* two corns, and so on. But this is to know that any three things are more than any two. What more do we know in knowing that three are more than two ? This only : that the amplitude of interests and desires being the measure of the world of things for every animate being, man knows more than does the crow in knowing that three are more than two, apart altogether from his ability to give symbolic expression to this knowledge. It is a difference of degree, not of kind.

Next, the more general assertion about qualities. I am not sure that I understand exactly what is meant by contemplating a quality apart from a thing. If it is meant that we can contemplate a quality apart from any image (whether visual or other) in which the quality is represented or typified, I must answer that, so far as I am concerned, I cannot do this. The image embodying the quality may be fleeting, unstable, fragmentary, but if there is no representative image at all, I am not contemplating a quality, though I may be contemplating a name. If the name evokes the recollection of a quality, I am contemplating that quality in an image which is, for the time being, typical. Make abstraction of all typical or representative imagery and nothing remains but the symbol and the knowledge that it is a symbol. If, on the other hand, by the ability to comtemplate quality apart from a thing, Whitney meant the capacity to see the general in the particular, which is the ability to form general ideas, then he was not consistent in his views, for he affirms in the most explicit way that some animals other than man are capable of forming general ideas :

'No one can successfully deny to the dog the possession of an intelligence which is real, even though limited by boundaries much narrower than those which shut in our own. . . . And anything wearing even the semblance of intelligence necessarily implies the power to form general ideas. It is little short of absurd to maintain, for instance, that the dog, and many another animal, does not fully apprehend the idea of a human being ; does not, whenever it sees a new individual of the class, recognize it as such, as having like qualities, and able to do like things, with other individuals of the same class whom it has seen before. . . . And how is any application of the results of past experience to the government of present action—such as the brutes are abundantly capable of—possible without the aid of general conception ? ' (p. 439).

That the *range* of mental action is in man far ampler than in other animals, rather than that there is any difference in the *mode* of action is, so far, all that we seem entitled to affirm. Let us next consider the validity, and the bearing upon our subject, of the statement that 'language-making is a naming of the properties of things, and of the things themselves through these properties'—an opinion which is explained and defended as follows:

'Let any one of us, even now, after all our long training in the expression of our conceptions, attempt to convey to another person his idea of some sensible thing, and he will inevitably find himself reviewing its distinctive qualities, and selecting those which he shall intimate, by such signs as he can make intelligible: there is no other way in which we can make a definition or description, whether for our own use or that of anybody else. If, for example, a dog is the subject of our effort, we compare our conception of him with those of other sensible objects, and note its specific differences—as his animality, shape, size, disposition, voice. This is so essentially a human procedure that we cannot conceive of the first makers of language as following any other. Then, in finding a designation, it would be impossible to include and body forth together the sum of observed qualities: in the first instance, not less than in all after time, some one among them would necessarily be made the ground of appellation. The sign produced would naturally vary with the instrumentality used to produce it, and the sense to which it was addressed: in the instance which we have supposed, if the means of communication were writing, it would probably be the outlined figure of a dog; if gesture, an imitation of some characteristic visible act like biting, or wagging the tail; if the voice, not less evidently an imitation of the audible act of barking: the dog's primal designation would be *bow-wow*, or something equivalent to it. But in this designation would be directly intimated the act; the actor would be suggested by implication merely: bow-wow, as name for "dog", would literally mean "the animal that bow-wows".'

All this appears to be as sound in sense as it is clear in expression; and we may remark in passing that it in no way commits us to the theory that language was solely onomatopoetic in its origin: a foundation assuredly far too slender to support the weight and bulk of the fabric subsequently erected. But perhaps we shall the better understand how the expression of thought, starting from the purely mimetic gesture, became what it is now and long has been, that is, conventional and symbolic, if, while not contesting the truth of Whitney's view, we consider an aspect of the matter on which he seems hardly to have dwelt with sufficient insistence: the intent and purpose of the namer. If, as in the instance contemplated by Whitney, the namer's purpose is to evoke the idea of a dog, and not merely to suggest by imitation that action of the dog which we call barking, ought we not then to say that, for him, *bow-wow* is the name of the dog, and is not the name, but is simply an imitation, of the action? Even in its origin, the essential function of a sign must frequently have been to evoke something other than, and different from, itself, something which, in experience, is associated with it; in other words, to suggest to the mind what is not depicted to the sense. The step from this to strictly conventional association, to the pure symbol as sign, is a long one; but it is facilitated just because, from the outset, the mimetic action was commonly intended on the one hand, and understood on the other, to direct attention to something beyond and besides that which it imitates. The proposition that qualities, properties, and actions in the abstract were the earliest subjects of naming is thus seen to be somewhat ambiguous. If 'naming', over and above what we commonly understand by the term, is also intended to cover 'imitation', then the proposition might be held to be true. But in what seems to be its literal sense, viz. that these abstractions, rather than the concrete experiences whence they derive, were what aboriginal man intended, by means of mimetic action, to converse about, it is surely not admissible; and it derives no support at all from what we know, or can surmise, of the habits of mind of savage tribes.[1]

Upon the more general question, How does language serve the process of thought in the individual, Whitney expresses himself as follows:

'In every department of thought, the mind derives from the possession of speech something of the same advantage, and in the same way, as in mathematical reasoning. The idea which has found its incarnation in a word becomes thereby a subject of clearer apprehension and more manageable uses: it can be turned over, compared, limited, placed in distinct connexion with other ideas; more than one mind, more than one generation

[1] Consult, on this point, Tylor's *Primitive Culture*, in numerous passages; Sayce's *Introduction to the Science of Language*, vol. ii, p. 4; also that less well-known but original work, E. J. Payne's *History of the New World called America*, vol. ii, p. 109.

of minds, can work at it, giving it shape, and relation, and significance. In every word is recorded the result of a mental process, of abstraction or of combination; which process, being thus recorded, can be taught along with its sign, or its result can be used as a step to something higher or deeper.'[1]

Again ('Darwinism and Language,' *North American Review*, July, 1874):

'thought is possible without language, but reflection is not; the thinker cannot hold up his thought before his own mental eye (at least, otherwise than in the most imperfect way) without the aid of symbols; words bring thought under the full review of consciousness.'

Finally:

'If we are pressed to say in what mode of action, more than in any other, lies that deficiency in the powers of the lower animals which puts language beyond their reach, we need have little hesitation in answering that it is the inferiority of the command which consciousness in them exercises over the mental operations: in their inability to hold up their conceptions before their own gaze, to trace out the steps of reasoning, to analyse and compare in a leisurely manner, separating qualities and relations from one another, so as to perceive that each is capable of distinct designation.'[2]

It seems to me there is here but very little advance on Hamilton. The same doubts and questions suggest themselves over again. We do not see *how* language does all this which it is supposed to do. Is it a fact that 'the idea which has found its incarnation in a word becomes thereby a subject of clearer apprehension and more manageable uses'? It may be so. It would certainly be rash to deny it; but it is not by any means self-evident. If I have an idea, and learn its name, it is certainly not obvious to me that I shall now more clearly apprehend the idea. If it is urged that the name, in its metaphorical applications, reveals the relations of the idea to other ideas, I would reply that this may be true, sometimes, of the way in which the individual acquires knowledge at second hand in learning language, but that this has nothing to do with the present question. It is clear that the *original* metaphorical use of a name waits upon the recognition of resemblance or analogy. While not contradicting the opinion that

[1] *Language and the Study of Language*, p. 419.
[2] Ibid., p. 440.

the incarnation of an idea in a word enables us to turn it over, to compare it, limit it, place it in distinct connexion with other ideas, we can affirm with far greater certitude that if we have an idea which is unnamed, it is the turning it over, the comparing it with other ideas, the discernment of its relations, of its differences from and its analogies with other ideas, which determine our choice of a name (by metaphor) for it from the available vocabulary, or of the existing materials for the coining of a new word. For we see this process actually at work, not only in the development of the terminology of the sciences, arts, and industries, but in the endeavours of young children to express ideas the names of which they have not yet learnt.

Whitney was on the right track in the opening sentence of the first of the above three excerpts, but he wandered off it. 'In every department of thought, the mind derives from the possession of speech something of the same advantage, and in the same way, as in mathematical reasoning.' There are two ways in which mathematical symbols aid mathematical reasoning. There is the particular way to which Whitney called attention: the artifice which underlies the naming of numbers, and which confers upon the process its real efficacy. It is clear that, whatever may be the mode in which words aid the reasoning process in general, it does not consist in this artifice. The other way, which Whitney barely noticed in passing, is familiar to us all: we can, and commonly do, use mathematical symbols in much the same way as we use counters in a game of cards; that is, with a knowledge of, but with no immediate attention to, their meaning. Now I believe that to some extent, but in a very much less marked manner, we can and do thus use words in general reasoning. Certain errors in reasoning, to which we are all liable, are readily explicable on this assumption that words are often used as mental counters. On the other hand, it is easy to see that such a use of words greatly expedites the process of reasoning, for it relieves us as much as we please, or as we dare to let it do so, of the burden of attending to the content of conception, and leaves us with the greater freedom to attend to the process of inference, or of deduction.

But it seems to have been denied, if not specifically, at least by implication, that words, as well as mathematical symbols, are used as mental counters. Some years ago, in the pages of

Mind, Professor Stout gave us the following neat definition: 'A word is an instrument for thinking about the meaning which it expresses; a substitute sign is a means of not thinking about the meaning which it symbolizes.'[1] This definition seems to have arrested attention and to have met with general acceptance. It is seductive, it is neatly antithetical, and it expresses with felicitous clearness the author's meaning. But is it quite true? It is an exaggeration. The distinction of function is drawn in terms so trenchant as to call for a mild protest: there is community of function as well. Be it remarked, first, that words, no less than substitute signs, are symbols, and that a substitute sign is a symbol which, usually for brevity's sake, is substituted for a number of other symbols. A mathematical symbol is a substitute sign in virtue of the definition we give of it in words; but the words are themselves symbols, and the intent of the definition is that the meaning attributed to the words is to be the meaning of the sign substituted for them. We must not let ourselves be led astray by the custom of language, which will have it that words express meaning while substitute signs symbolize it. It is as well to remember that it is we who express words and other signs, while the words and other signs symbolize what we mean.

The distinction is one of degree rather than of kind. In our use of substitute signs we avail ourselves of them more as mental counters, and less as symbols whose meanings need meditation, than we do in the case of words. But, at bottom, what governs the use we make of either, and in either capacity, is the nature of the conceptual system which is the subject of symbolization. Where that system is clear, definite, and stable, that is, where the work of conception is, in the mind of the conceiver, complete, the symbol naturally and advantageously lends itself to use as a mental counter. That words, as well as substitute signs, are very frequently used as mental counters is hardly open to question. I pass by the evidence afforded by self-introspection, because the habit of self-introspection, with reference to purely intellectual processes, is rare, and is of little use when not carried out methodically. But, to return to a point already alluded to, it is a matter of common knowledge, and within every one's experience, that in the course of reasoning on any subject we are all liable to be betrayed into false or ambiguous conclusions

[1] *Mind*, April, 1891.

through inadvertently using a term or a phrase in more than one sense. Now if it were true that in deliberate reasoning we always use the word as a means of thinking of the meaning which it expresses, and never as a mental counter, the frequency with which this fault is committed, even by careful thinkers, would be inexplicable. The inadvertence arises from failing recollection of the sense attributed to the word at the outset. But it is obvious that if this meaning were fully recalled every time the word is used in the course of the argument, the probability of an unintentional modification of meaning would be practically nil. On the other hand, if in the intermediate steps of the argument the word is used more or less as a counter, with at most but a very perfunctory recall of the meaning, the frequency with which an unintentional change of meaning creeps into a train of reasoning is readily comprehensible.

If it can be granted that, in general reasoning, we employ words to some extent as we use substitute signs in mathematical reasoning, then it is granted that the use of words in the process of reasoning expedites that process: there is economy of mental effort. But, as we have seen, much more than this is claimed for language as an instrument of thought. It is claimed that it enlarges the scope of the reasoning process and lends stability to it. There is obviously a sense in which it may be said that language enlarges the scope of the process of reasoning when it is admitted that it expedites that process. But this is not what is meant here. The claim is that in the thinker's use of words the field of conception is broadened, thus providing a greater wealth of material for the exercise of the process or, at the least, rendering that material more readily available. So far, however, we have been unable to see how the use of words in thinking can effect anything like this. Methodical observation and experiment sometimes widen the field of conception because they lead to the discovery of resemblances and analogies on the one hand, and of distinctions on the other, both of which, the one directly, and the other indirectly, involve extension of acquired concepts or acquisition of new ones. Reflection upon, and comparison of, acquired concepts will also at times issue in the discernment of differences and similarities previously unnoticed. If we assume, with Whitney, that the ability to reflect is dependent upon words, this no doubt settles the question, but it is to abandon any attempt at explanation, and at the same time it obliges us to ignore the emphatic assertion of competent witnesses that they 'think less easily in words than otherwise'. On the other hand, if we accept the latter statement without qualification, as an adequate expression of the experience of those who make it, we must then conclude either (1) that the value of language as an instrument in the process of thought is simply a matter of personal idiosyncrasy, or (2) that the common belief that subjective verbal formulation is a help in, if not a necessary condition of, carrying out this process, is an illusion, the origin of which is to be found in the difficulty we experience when we endeavour to dissociate language from the process.

I am quite unable to accept either of these conclusions, and I suppose hardly any one who has thought over the matter would be prepared to do so. The conflict of opinion is no doubt real, but, as so often happens in other disputes, it is partly due to lack of mutual understanding, to the employment on both sides of the same general terms without sufficiently close agreement in the meanings attributed to them. Some years ago an interesting discussion on the subject appeared in the pages of *Nature*, among the disputants being Max Müller, Romanes, and Francis Galton. A critical estimate of this discussion will, I think, enable us to ascertain the real facts, and thus put us in the way to arrive at a definite and satisfactory conclusion. The discussion arose from the publication in 1887 of Max Müller's *Science of Thought*, followed by his delivery of three lectures at the Royal Institute on the same subject. The entire correspondence, as it appeared in *Nature*, is reprinted as an appendix to the said lectures.[1] The main point at issue can be made sufficiently plain by a few brief extracts from the letters of Galton and Max Müller.

Galton, in his first letter, writes:

> 'It happens that I take pleasure in mechanical contrivances; the simpler of these are thought out by me absolutely without the use of any mental words.'

He also gives as examples the thinking out of strokes in billiards, and the calculation of moves in chess, by eye alone.

> 'In simple algebra, I never use mental words. . . . In simple geometry I always work with actual or mental lines; in fact, I fail to arrive at the full conviction that a problem is fairly taken in

[1] *Three Introductory Lectures on the Science of Thought*, 1888.

by me, unless I have contrived somehow to disembarrass it of words. It is a serious drawback to me in writing, and still more in explaining myself, that I do not so easily think in words as otherwise. Professor Max Müller . . . has fallen into the common error of others not long since, but which I hoped had now become obsolete, of believing that the minds of every one else are like one's own.'

All these instances which Galton gives of reasoning without words, and of the oppression of words in reasoning, are, with one exception, instances for him—whatever they may be for any one else—of reasoning by means of representative or typical images, and, whether symbolic images or words accompany the process or not, it is easy to understand that when they do, the reasoner should find them useless and even burdensome. The one exception to which I allude is algebra. Galton spoilt his case by including this instance among the others, and Max Müller at once fastened on the weak point, urging with unquestionable force that ' in algebra we are dealing not only with words, but with words of words. . . .' The fact is, of course, that algebra is a mode of reasoning almost purely symbolic ; and although Galton was no doubt formally justified in the bare statement that he did not use *words* in algebraic reasoning, it argues a curious blindness to the fundamental question involved in the discussion that he should have put simple algebra side by side with simple geometry ; the latter being a subject for which, in the actual ratiocinative process, symbols are wholly superfluous. It may be added, too, that the remark about believing the minds of every one else to be like one's own, though it emphasizes a caution which should not be neglected, was here somewhat out of place : as Max Müller justly contended in his reply, ' the identity of language and reason can hardly be treated as a matter of idiosyncrasy.'

But, as against the genuine instances of unsymbolic thinking adduced by Galton, what sort of arguments did Max Müller urge ? Galton says in one of his letters :

'There are street improvements in progress hereabouts. I set myself to think, by mental picture only, whether the pulling down of a certain tobacconist's shop . . . would afford a good opening for a needed thoroughfare. Now, on first perceiving the image, it was associated with a mental perception of the *smell* of the shop. I inhibited that mental smell because it had nothing to do with what I wanted to think out. So words often arise in my own mind merely through association with what I am thinking about; they are *not* the things that my mind is dealing with; they are superfluous and they are embarrassments, so I inhibit them.'

To this Max Müller answers: 'You say . . . you inhibit any mental word from presenting itself. What does that mean if not that the mental words are there.' But of course no one disputes, or could dispute, the fact which Max Müller here irrelevantly asserts. Is it necessary to point out that the question is whether, in the given instance, the words which admittedly are present are unessential adjuncts to the process of reasoning, or necessary instruments in that process? Again, with regard to chess-playing, Max Müller argues that 'a chess-player may be very silent, but he deals all the time with thought-words or word-thoughts. How could it be otherwise?' This is mere reiteration. If chess-players are mistaken in supposing that they play entirely by means of visual imagery, they will never be convinced of their mistake by such an argument; nor by the one which follows: 'I thought that to move a castle according to the character and rules originally assigned to it was to deal with a word-thought or thought-word. What is a "castle" in chess, if not a thought-word or word-thought? I did not use the verb "to deal" in the sense of pronouncing, or rehearsing, or defining, but of handling or moving according to understood rules.' But what is asserted is that chess-players can, whether in act or in imagination, move a 'castle' according to assigned character and understood rule without any need for its name or for any verbal formulation, silent or other, of character and rule.

Finally, being hard pressed to justify his general conclusion, Max Müller is driven to the too common expedient of slipping it by definition into the premisses of his argument: 'In the same manner, when Professor Romanes takes me to task because I said that no one truly thinks who does not speak, and that no one truly speaks who does not think, he had only to lay the accent on *truly*, and he would have understood what I meant—namely, that in the true sense of these words, as defined by myself, no one thinks who does not directly or indirectly speak, and that no one can be said to speak who does not at the same time think.' Every one, it has often been said, is at liberty to define the sense in which he uses particular terms, but more certainly still is every one free to ignore a definition which begs a point at issue.

I have drawn attention to this debate in *Nature* because it helps to make plainer still what was already, I should imagine, tolerably plain: that the value of language or of symbolism of any kind in the prosecution of the thought-process very largely depends upon the content of that process, that is to say, upon the nature of the subject thought about. But, if I am not mistaken, it does very much more than this. It puts us on the track of the main cause of the efficacy of the symbol in the mental development of the individual, irrespective of the advantage we may derive from it when judiciously used as a labour-saving device.

Max Müller's doctrine of the essential unity of thought and language, as we saw in the first chapter, is irreconcilable with the modern belief in the continuity of development. To admit it is to admit the existence of a sheer abyss between mental action in man and mental action in other animals. Nay, more than this, it is to deprive the expression 'mental action', when predicated of the other animals, of any intelligible meaning. We can then hardly avoid taking the retrograde step to Descartes, and declare the brutes to be automatons; for if analogy of the brutes' actions with man's is insecure ground for the belief that they think, the belief that they feel is no better founded.

Perception, according to Professor Binet, is itself a rudimentary process of reasoning.[1] Even if we are unprepared to admit the proposition on the evidence afforded by experiments in hypnosis and other hallucinatory states, we can scarcely read the analyses given by psychologists of the process of normal perception without discerning the close analogy between this process and that which goes by the name of reasoning. Symbolic images forming no part of the brutes' process of thought we cannot conceive this process in the animal, otherwise than as embodied in representative images; but we can do little more than guess in a vague way at the extent, the fullness of conception which is involved in the perceiving even of the higher animals; it is, in other words, a matter of conjecture under how many aspects an image is recognized by the animal, to what extent it may be representative or typical. The greater we admit the difference to be between

[1] Alfred Binet, *The Psychology of Reasoning*, chap. iii.

the man's and the brute's ability to discern similarities and dissimilarities, analogies and distinctions, the more we circumscribe the field of conception in the brute's mind by comparison with man's.

But now let us make abstraction of this difference in innate capacity of discernment, and consider merely the fact that man associates symbolic with representative images while the brute does not. If we reflect that a symbolic image, a conventional sign, a name, has no sort of likeness to the things or mental images with which it is associated, a sudden flood of light is thrown upon the subject, the enormous importance of the symbol breaks in upon us, and we see in the fact of such association not, indeed, the secret of the greater powers of thought in the man—for this secret lies hidden in the unknown differences of cerebral organization—but the principal means by which man's thinking gains amplitude in its range and stability in its results. For must not this association of names or of purely symbolic images of any kind with representative images tend to counteract a tendency inevitable where the process of thought is embodied solely in representative imagery: the tendency to mental reinstatement of less general concepts at the expense of more general ones? Resemblance is not only of many kinds, but of varying degree in each kind; remote likenesses are less readily cognized, arrest attention less, than those that are more immediate. The recognition of a thing or of a mental image is the awareness of its similarity to other things or images. Since this awareness is the more vivid the closer is the degree of likeness, it follows that the concepts involved in and represented by the images in a train of mental imagery must commonly possess but a low degree of generality—it is not that there need be incapacity to discern remote likenesses or analogies; the point is that the mechanism of recall or reinstatement diminishes in effectiveness with the increase in degree of generality of the resemblances or analogies which may have been discerned. We may put it that representative imagery, when not associated with symbols, tends to limit the scope of inference because it hinders the ready availability of general concepts as ground or material for inference.

How does the association of symbolic with representative images tend to counteract this cramping effect; how does it help to remove the obstacle to easy reinstatement of general concepts?

The answer is plain: because the association is artificial; because the connexion of the same name with each one of a series of mental images, themselves associated in virtue of some common likeness, is an artificial connexion. Of a number of images which, while differing from one another in varying degree, have been classed together under a common name because of a common likeness, no one is any more or less like the name than is any other, no one has any closer natural connexion with it than has any other. If an image evokes a common name, the name may in its turn as readily recall any one as any other of the more or less like images with which it has been associated as a name, while the tendency of the (representative) image must be to recall those more closely associated with it in likeness.

I perceive or recollect an object. There is something, an element of the percept or image which, in my present apperceptive mood, specially arrests my attention. This element is in my mind associated with, and evokes, a name: the word 'green'. But it is also associated with and may evoke another name: the word 'colour'. This word has been associated with, has evoked and has been evoked by a number of other elements (in other percepts and images) between which I have discerned a certain definite similarity. We may say that this word 'colour' is *the* name of the concept involved in the discernment of this similarity, while it is *a* name of every one of the elements subsumed under the concept. Again, I perceive or recollect an object which is associated in my mind with the word 'horse'. But it is also associated in my mind with such terms as 'animal', 'herbivore,' 'mammal,' 'domestic.' It does not follow, of course, that when I perceive or recollect a horse all or any one of these other names will be evoked, but the fact that they are all associated with the percept or image, and *may* be recalled by it, constitutes a means of reinstating concepts which is lacking in the mind in which an unsymbolic image is the sole link of recall for the various concepts under which it may have been at various times subsumed. It is a means of making the process of thought more discursive, and this, when the means is under the control of the thinker, is invaluable. It is obviously one of the most important conditions of mental progress.

We may, then, define one of the essential functions of words in the process of thought to be the readier recall of those concepts which are not immediately involved in ordinary recognition. In the constant recall through verbal association of the more general concepts and abstractions they become permanent, instead of fugitive possessions. It is upon the permanence, the familiarity with, general concepts and abstractions that depends the possibility of general inferences; one thus realizes vividly enough how words, as instruments for thinking of the meanings which they express, enormously enlarge the scope of the ratiocinative process: and we have already seen how, as substitute signs, they may expedite the conduct of the process. It will also be clear that the character and extent of these two aids to reasoning, their effective application, must depend largely upon the nature of the subject of reasoning, that is, upon the nature of the conceptions involved in that subject; that in some subjects these aids may altogether vanish; that words may become superfluous, may even become embarrassing, in the conduct of the process.

This sketch of the relations of thought and language was written now a good many years ago. The subject is evidently one proper to the psychologist, but it is only within the last decade, so far as I know, that it has received from psychologists the attention it deserves. In this connexion I should like to refer to Professor Stout's *Analytic Psychology* and *Manual of Psychology*. A careful perusal of these two works showed me the necessity of making some corrections in the original draft of my sketch, of toning down here and there an opinion too trenchantly expressed, and of bringing the terminology into closer accord with that now generally current in psychological analysis. But in its essential features I have left it unaltered, and I do not think there is any point on which I may not agree with that author, which is at the same time a point of vital importance in relation to the subjects discussed in the rest of this work.

It is true that Dr. Stout holds, generally, that conception, save of the most rudimentary kind, is not possible without language; but if I understand him, this opinion is not pressed absolutely, without qualification and attenuation, so that I do not know how far I may be in real opposition to him in regarding the ability to form concepts as a precondition both of the invention and the learning of language.[1] But real opposition there certainly is if it is affirmed that awareness of resemblance between particulars is not properly speaking a concept, or 'cognition of a universal as distinguished from the particulars which it unifies';[2] and if it is further affirmed that this kind of cognition is possible only through the agency of language. I do not see my way to come round to this opinion. My difficulty as to the first point is that I can find no real distinction between the cognition of a universal and the apprehension of the kind of resemblance which connects the particulars. Special stress is laid, in the explanation, on the qualifying words 'as distinguished from'. I distinguish the universal from the particulars which it unifies, but this means to me no more than that I attend to a common feature by contrast with the differences which constitute the individuality of the particulars. It is easy to say that this synthetic-analytic process is one 'which probably cannot take place except in a very rudimentary form without the aid of language', but is there any good reason for supposing that relative ability to form concepts depends upon anything else than relative acuteness of discernment and interest in surroundings, that is, upon relative development of the brain? Familiar and effective possession of concepts sporadically formed is another matter.

Words, as instrumental in the process of thought, have in my view two distinct functions: (1) they accelerate the process when we use them as substitute signs; (2) they broaden the basis of effective conception in their vast superiority to representative imagery for the embodiment of conceptions of a high degree of generality, in addition to which they must also be more effective than representative imagery in maintaining definite distinction between very similar or closely allied concepts. This second function depends upon the very nature of the symbol as something wholly unlike, and in no natural way connected with, that which it symbolizes. There is, I think, a fairly close connexion between this view and what Dr. Stout says about the shortcomings of gesture-language:

'The primary and essential procedure of the language of

[1] Cf. *Analytic Psychology*, vol. ii, chap. ix.

[2] *Dictionary of Philosophy and Psychology*, vol. i, p. 208. Professors Stout and Baldwin are jointly responsible for the definition and the explanation which follows it.

natural signs is to represent things and processes by imitating the broad features of their sensible appearance and especially of their appearance to the eye. But the characters which are capable of being so imitated are of a comparatively low grade of generality or abstractness. They represent an analysis of perceptual experience into universals and its reconstruction out of these universals. But the universals themselves are very far from being simple and ultimate. They in their turn are intrinsically susceptible of analysis, and constituents thus revealed are again susceptible of further analysis, and so on. Now the more advanced is this process of dissection, the more helpless is pictorial representation to express the result either within the individual consciousness or in the intercourse of different minds. But the power of mentally representing a universal is simply the power of conceiving it. Where the mental representation does not exist, the conception does not exist. Hence a mind whose discursive thinking could only find expression in self-interpreting signs, would be incapable of the higher reaches of abstraction.[1] Broadly speaking, natural signs are capable of fixing attention on universals which are constitutive characters of particular objects as presented in perceptual experience ; but they can only to a very limited extent fix attention on universals which are constitutive characters of other universals. The thinking which depends on the imitative gesture generates concepts ; but it can hardly generate a conceptual system, in which there is an ascending scale of generalization, passing from species to genus, and from genus to higher genus, and so on through a series of gradations till the highest genus is reached. It seems beyond the unaided powers of the thought which works through natural signs to frame a system of classification.'[2]

Substitute, for natural signs, representative imagery (and all non-symbolic imagery, in so far as recognized, is necessarily representative), whether susceptible or not of 'natural' expression, and much that is said in the above passage as to the limitations of natural signs for the individual consciousness applies to representative imagery. And I doubt not that Professor Stout

[1] I do not think this necessarily follows. Why should not the power of conceiving outrun the range of self-interpreting signs ; and is it not in the highest degree probable that the development of convention in expression was due mainly to such a condition of things ? Given the invention of an artificial sign for a hitherto unexpressed and unusual general conception, and it will become a permanent intellectual possession, partly because it will no longer be ousted by the less general conceptions more immediately involved in representative imagery, partly because it becomes a possible subject of intercommunication.

[2] *Manual of Psychology*, 2nd edition, pp. 504, 505.

would in substance agree with the view that the value of the artificial sign in the expanding process of conception lies precisely in the fact that it is not representative, but symbolic.

I conclude with a reference to the footnote on page 19. In there laying stress on the necessity of admitting both the objective and the subjective aspects of language, I had in mind those two flatly contradictory theories of the relations of thought and language of which Max Müller and Whitney were respectively exponents. For Max Müller, thought and language being essentially one, language could be nothing but subjective. According to Whitney, language is, on the contrary, essentially objective :

'The essential determining element in any real theory of the beginnings of speech must be the same recognition of the true nature of speech which we have already found efficient above in deciding other questions : namely, that words are only external signs for conception.' [1]

If we disengage each of these opposed views from the erroneous implications involved in it, they become complementary. In so far as Max Müller's exposition is to be taken as a justification of the literal identity, the 'essential oneness' of name and concept, sign and signification, word and meaning, it cannot stand : we can no more shut our eyes to the enigmatic nature of the assertion than to the incongruities and obscurities which abound in the endeavour to substantiate it. Yet when we reflect upon the insistence with which to the very end, and notwithstanding the most damaging criticism, Max Müller steadfastly upheld his theory, we cannot help suspecting that there must have been throughout in his mind the persistent consciousness of some essential fact which his opponents neglected. And, indeed, if we must brush aside that impossible identity for which Max Müller strove, and that vain attempt to establish an absolute discontinuity between the thinking of animals and that of man, this does not prevent us from admitting that unity of language and thought which is the clear recognition of the fact that a word, in its subjective aspect, is an element of thought. Moreover, although we cannot assent to the unqualified statement that there is 'no language without reason, and no reason without language', we may allow it to pass in so far as it gives expression to the highly probable

[1] *Max Müller and the Science of Language.*

fact that the ratiocinative process in man, when it involves conception of a high degree of generality, or of nice gradation, can be effectively conducted only by the use of symbols.

Turning now to Whitney's views of the true nature of speech, 'namely, that words are only external signs for conceptions,' we cannot but demur here also. The sciences of phonetics, of phonology, of inscriptions, deal with the external signs for conceptions. Yet any one may be deeply versed in any of these branches of knowledge, or in all of them, without ever having troubled himself to form an opinion on the point about which Max Müller and Whitney so completely disagreed; on the other hand, a man may be almost wholly ignorant of these sciences and yet be competent to express a reasoned opinion on the matter in dispute. Consider it, if you please, an exaggeration, or even a misuse of language, to say that these external signs are words only by metaphor; such exaggerations or verbal misuses are sometimes necessary to startle us out of our habitual grooves of thought. It is impossible to suppose that a man of Whitney's capacity and penetration should not have been perfectly aware, or should have intended to deny, that there is a subjective side to language; but he evidently considered this to be of little or no importance for the right understanding of the relations of thought and language. We must take his repeated insistence on the externality of signs rather as a repeated protest against the untenable theory of the identity of thought and language in the sense maintained by Max Müller. But the momentum of his protest carried him too far. It is useless and even misleading to ignore that language is an integral part of thought: to affirm it is merely to lay the necessary stress on its subjective aspect.

PART II

IMAGINARY QUANTITIES IN ALGEBRA

AND

IMAGINARY LOCI IN GEOMETRY

CHAPTER IV

ON THE NATURE OF THE CONCEPTS OF NUMBER, QUANTITY, MAGNITUDE, AND MEASURE

Nature of the concept of Number.—Its independence of, but close association with, the concept of Order.—Interdependence of Number-concepts.—Number-concepts as represented and as symbolized.—Transition from the representative to the symbolic image.—Distinction between Number and Quantity in applied Mathematics.—Stallo's criticism of the use of the term quantity in connexion with Algebraic symbols.—Use of the terms Number and Quantity in pure Algebra.

LET us suppose an intelligent child to know the meaning of the word 'many', and of the word 'same', but not yet to have learnt to count: what might we suppose the child to be thinking about if he asked us the question: What is it called when I think of a same many? We should probably surmise that he was thinking of some number, that he expected that numbers must, like everything else, have names, and that he wished to know the name of the number he was thinking of. But if, on spreading out before him a few coins or other small objects and asking him to show us which 'same many' he meant, he pushed the things aside and said 'any same many', we might indeed be astonished at this evidence of interest in abstract thought, but we should be obliged to conclude that the child had very definitely formed the abstract conception to which we give the general name 'number', and that he wished to know that name.

The supposition no doubt implies a greater power of abstraction, as well as a higher degree of purposiveness in thought, than are at all likely to characterize any child in the earlier stage of learning language: I make use of it as illustrative, not as probable, as exemplifying what appears to me to be the essential trait in the conceiving of number, apart from the artifices involved in counting.

Number-concepts, like all abstractions, originate in the innate capacity to discern likeness and difference and to concentrate attention on some one element of a complex presentation. But while the vast majority of familiar concepts arise in the discernment of approximate likenesses and differences between individual things, the concepts of number spring from the discernment of exact likeness and difference between individual aggregates of things, merely as aggregates. The legs of a horse, of a cow, of a dog, of the chair upon which I am at present sitting; the limbs of my body; the fingers (excluding the thumb) of my hand; the sides of a square; the corners of a square: in these things, taken individually and compared one with another, I discern various kinds and degrees of likeness and difference; but taken in their several aggregates and compared merely as aggregates, I discern a certain exact likeness, or identity (the 'same many' of the child), which is the concept or abstraction to which I have been taught to give the name 'four', and have learnt to associate with the mathematical symbol '4'.

The discernment of identity involves that of difference. It is impossible to have become aware of identity of aggregation without having *pari passu* become aware of that difference of aggregation which we indicate by the terms 'more' and 'less'. The perception of non-identity may be merely that of 'more or less difference', that is, may be indefinite. But it is impossible to have become aware of the identity between the legs of a dog and those of a horse as aggregates, between either of these and the sides of a square, between any of these and the corners of a square, &c., without having become aware that the difference between this identity and that of the hind legs and tail of a dog and those of a horse, of either of these and the sides of a triangle, of any of these and the corners of a triangle, &c., is in every case the *same* difference, that is, is itself an identity, the identity called 'one', the name given to the individual as part of any aggregate.[1]

[1] The interdependence of number-concepts is partly reflected on the symbolic side by the interdependence of number symbols or names in definition. No significant definition of the name or symbol of a number can be given without the aid of other number symbols or names whose meanings are already known; and in this as in every other case, definition ultimately abuts upon some mode of conveying meaning which is not definition, upon an appeal of some kind to the intuitional and non-conventional in personal intercommunication. In the opinion of recent investigators in the philosophy of mathematics, however, numbers (cardinal) can be defined each independently of all the others, and without assigning to them any order or relation (cf. Couturat, *Les Principes des Mathématiques*, 1905, p. 52)—a surprising feat if the conceiving of numbers

This rudimentary conceiving of numbers, or progressive discernment of identity and difference between aggregates of familiar things, is of course wholly independent of any names or other symbols for numbers ; it is probably independent, for a considerable time, of systematic representative imagery such as that suggested by the fingers and toes. To the consciousness of the mature individual this discernment, in the case of small aggregates such as those above mentioned, appears immediate or intuitive ; but we must admit the probability, if not the certainty, that for the immature individual it was the result of an oft-repeated and laborious mental process : the one-to-one co-ordination of the things in one aggregate with those in another.

Every such co-ordinative process, to be effective, necessitates attention to the order of the co-ordination : we could not otherwise recollect which individuals of two groups have been, and which yet remain to be, co-ordinated—save in the case where the co-ordination can be carried out manually. It is clear, then, that from the outset the concepts of number and of order get closely associated and tend to develop together. So intimate does this association become in the mind of the civilized thinker, habituated from childhood to the serial association of names for numbers, that many have endeavoured to derive the concept of number from that of order. Cayley, for instance, following Helmholtz, Kronecker, and Dedekind, regards cardinal numbers as derived from ordinal numbers.[1]

Again, it is this close relationship between the concepts of order and number which suggested the artifice characteristic of

involves the conceiving of identity, and hence also of non-identity, between aggregates. The definition in question runs thus : the number of a class is the class of all classes similar to the given class (Russell, *Principles of Mathematics*, 1903, p. 115). In order to understand it, read it thus : the number of a class (i.e. aggregate) is the class (i.e. similarity) of all classes (i.e. aggregates) similar to the given class (i.e. aggregate) ; the term 'similar' being used in the sense of 'identical' as I have used it above. The definition has, for me at least, no meaning other than this ; and this would certainly not in general be admitted as defining *the* number of a given class or aggregate. But then, for Mr. Russell, the 'philosophical sense' of definition ('analysis of an idea into its constituents') is useless, at all events in mathematics (cf. pp. 111, 112). It will be seen that this definition of the number of a class in no way invalidates what I have said above as to the definition of the name or symbol of a number.

[1] Cayley, *Collected Mathematical Papers*, v. 292–4 ; xi. 442–3.

the invention of names for numbers, viz. the memorial association of the names in fixed serial order; and as this order is, in the nature of the case, an order in time, it has been argued by other thinkers—among mathematicians by Sir W. R. Hamilton, the inventor of Quaternions—that the concept of number is derived from that of time.

The fact that we are taught to count by the aid of names from a very early age, and thus learn to think about numbers by means of symbolic instead of representative imagery; and the equally if not more important fact that we retain in memory little or no trace of the mental processes through which we originally elaborated the simple, fundamental, primary conceptions which are the common property of all men—these facts tend in many minds, and not least in the most educated, towards a practical obliteration of the distinction between numbers and the symbols, verbal or mathematical, with which we have been taught to associate them. Neither names nor mathematical symbols, however, are necessary in order to form concepts of number, nor are they indispensable for the effective and ready use of these concepts in the process of reasoning, so long as the process does not overpass a certain degree of complexity. Images, representative or typical of number-concepts, are, up to a certain point, fully adequate for the purpose; and, naturally and universally, the fingers have become such typical images, for they accompany us whithersoever we go as the most familiar of group-images, as the most familiar instance of fixed serial order, and of an ordered group which lends itself to significant repetition.

This is not a theoretical deduction, but an observed fact, so far as any mental fact can be said to be observed. No one who has read with attention such works as Tylor's *Primitive Culture*, or Professor Conant's *The Number-Concept*, can, I suppose, entertain the least doubt about the matter. These works, and others which cover the same field of inquiry, overflow with evidence that among savage tribes the absence of names or other symbols for more than the first few numbers is no obstacle to the conceiving and expressing, by representative imagery, of numbers far in excess of those for which they have names. Surprise is sometimes felt that certain savage tribes show a reluctance to use such names for numbers as they possess, and resort in preference to finger pantomime. But it is not difficult to understand, given a stage of civilization in which the need for thinking of and expressing relatively large numbers does not often arise, that names for such numbers, although actual tribal possessions, should be felt as a burden on the memory by the great majority of the individuals constituting it. The fact is that to the civilized individual, saturated from childhood with the symbolism of arithmetic, the conceiving of numbers comes to be regarded as one with the use of numerical symbols, so that he sees, in the savage's pantomimic expression of numbers and lack of names for them, evidence of his incapacity to frame abstract numerical concepts. I will illustrate my meaning by a passage from the above-mentioned work of Professor Conant. It should be perused with close attention to the senses in which the terms 'concept', 'image', 'abstract', 'concrete', are employed.

'In the sense in which the word is defined by mathematicians, *number* is a pure, abstract concept. But a moment's reflection will show that, as it originates among savage races, number is, and from the limitations of their intellect must be, entirely concrete. An abstract conception is something quite foreign to the essentially primitive mind, as missionaries and explorers have found to their chagrin. The savage can form no mental concept of what civilized man means by such a word as 'soul'; nor would his idea of the abstract number 5 be much clearer. When he says five, he uses, in many cases at least, the same word that serves him when he wishes to say *hand*; and his mental concept when he says *five* is of a hand. The concrete idea of a closed fist or an open hand with outstretched fingers is what is uppermost in his mind. . . . He sees in his mental picture only the real, material image, and his only comprehension of the number is, "these objects are as many as the fingers on my hand."'[1]

What Professor Conant means is, I suppose, this: that when a savage thinks of the number five, he thinks this concept under a quite different kind of image from that under which a civilized man thinks it—unless, indeed, he means that the latter does not think it under any image at all, because a word or other symbol is not commonly classed as a mental image. This, however, is merely a matter of phraseology, not in the least affecting the fact that both the civilized man and the savage think the concept in question under some mental presentation. The difference, then, lies in this, that the savage thinks the number five under an

[1] Conant, *The Number-Concept*, p. 72.

image which is representative or typical of the concept, while the civilized man thinks it under an image which is symbolic of it. We may say that the one way is artificial, the other natural, in the sense in which it would be allowed that the artificial is more characteristic of civilization than of savagery; but the concept, in whatever customary image embodied, is just that concept, neither more abstract, nor more concrete, nor more 'pure' in the one case than the other. And if the image which is present to the savage mind may properly be called concrete, I see no reason why the image present to the mind of the civilized man should not be likewise qualified. In fine, it is the concept which is abstract, and the image which is concrete; the real difference lies in the nature of the image under which the concept is thought.

A savage is asked some question, the answer to which involves the number sixteen. He may or may not be able to conceive this number: that will depend partly on the general average of intelligence of the tribe to which he belongs, partly upon whether the conditions of life and the nature of the customs of the tribe are such as to promote the endeavour to conceive numbers; e.g. the custom of bartering. But we will suppose him to answer, say by extending the ten fingers, then again the fingers of one hand, together with one finger of the other. Term this process pantomimic if you please, it is none the less reasoning and inter-communication by means of representative imagery. The conceptual counterpart of this pantomimic expression, in the mind of this untutored savage, is not essentially different from what mine would be were I to convey the same information by writing the equation $x = 10 + 5 + 1$, i.e. the number unknown to you is identical with the sum of 10, 5, and 1. Although the savage may have no name for the number, he clearly conceives it as an aggregate of other aggregates, each of which is represented by a distinct and familiar group-image.

Any system of conceiving and expressing numbers, such as this, in which the imagery remains purely representative, is obviously limited by the increasing burden it imposes on the memory. As the necessity for expressing higher numbers makes itself felt, artifices for lightening the burden will be sought. It seems natural to suppose that the transition from the representative to the symbolic image will be gradual, that there will be a stage at which, together with images which are purely representative, there will also be images which can be classed neither as purely representative nor as purely symbolic. Archaic systems of numerals, or inscribed signs for numbers, all show a combination of the representative with the symbolic image; and evidence of transitional images in the reckoning process of some savage tribes is not wanting. Here is an example:

'More than a century ago travellers in Madagascar observed a curious but simple mode of ascertaining the number of soldiers in an army. Each soldier was made to go through a passage in the presence of the principal chiefs; and as he went through, a pebble was dropped on the ground. This continued until a heap of 10 was obtained, when one was set aside and a new heap begun. Upon the completion of 10 heaps, a pebble was set aside to indicate 100; and so on until the entire army had been numbered.'[1]

Here the successively formed heaps of ten pebbles are purely representative, but we can hardly assign to the pebbles successively set aside both for tens and for hundreds, either a purely representative or a purely symbolic character. In so far as one pebble is made to stand for ten pebbles, or ten soldiers, there is a beginning of symbolism; but there is still a strong element of representation in the substitution of one pebble for one heap, two pebbles for two heaps, and so on.

The use of the fingers to express numbers, a use which appears to have been universal, and in its origin always representative, passed over in the old world into a use almost entirely symbolic, and said to have survived in the East to the present day. The original sources of information on this subject are given in the article 'Numerals', *Encyclopaedia Britannica*, vol. xvii, p. 625, footnote. The following brief description of the system is taken from the article itself:

'In the later times of antiquity the finger symbols were developed into a system capable of expressing all numbers below 10,000. The left hand was held up flat with the fingers together. The units from 1 to 9 were expressed by various positions of the third, fourth, and fifth fingers alone, one or more of these being either closed on the palm or simply bent at the middle joint, according to the number meant. The thumb and the index were thus left free to express the tens by a variety of relative positions,

[1] *The Number-Concept*, p. 8.

e.g. for thirty their points were brought together and stretched forward; for fifty the thumb was bent like the Greek Γ and brought against the ball of the index. The same set of signs, if executed with the thumb and index of the right hand, meant hundreds instead of tens, and the unit signs, if performed on the right hand, meant thousands.'

If we compare this gesticular system with others, articulate or written, which are familiar to us, we see that the artifices of expression resorted to are in all of them as nearly identical as the different natures of the medium of expression allow. Be the symbols articulate sounds, gestures, or written marks, they are in all three cases subjected to analogous operations which are symbolic of the conceptual process. Apart from the meaning assigned to each symbol as a separate entity, symbolic operation, i.e. the putting of the symbols into context, temporal or spatial, determines by rule or custom the meaning of that symbol in that particular context. In the naming of numbers, as compared with the naming of other conceptions, the distinguishing trait, as we have already remarked, is the fixed serial association of the names in memory. But, comparing the name-system with other symbolic systems in aid of conceiving and expressing numbers, it is seen that serial association of the symbols is not a peculiarity of the name-system; it is characteristic of them all. Obviously, neither arithmetical symbols nor the system of finger symbols above described would retain their significance apart from their being remembered as a series. This is not inconsistent with the opinion that the initial conceiving of numbers as identities in aggregation is independent of the conception of fixed serial order, either of things in space or of acts in time; it is as an effective aid in establishing these identities that the conception of fixed serial order is indispensable. A series of number-symbols is thus, one might say, a reference series of individuals, each of which has a mark which assigns to it its place in the series, only the individuals have vanished and left their place-marks behind. A number-symbol may thus be considered and intended, according to the purpose of our thought, either as the symbol of an identity in aggregation without regard to order, or as that of a place in a serial order of places, temporal or spatial.

In the exposition of mathematical thought the terms Number,

Quantity, Magnitude, and Measure, meet us at every turn. But while, in applied mathematics, writers who avoid looseness of terminology are careful to indicate, either by definition or by clear implication and example, the precise meaning which they attach to these terms, in pure mathematics it is a common if not invariable custom for writers to use these terms loosely, without any clear intimation of the shades of meaning intended, if any are intended. Thus, in textbooks of Algebra, the term 'quantity' is often suddenly introduced, without a word of explanation, into expressions where it is apparently intended to replace the word 'number' as a synonym. But as neither in the ordinary nor in the philosophical use of these terms are they customarily intended as synonyms, we can feel no security, in the absence of explanation, as to the writer's real intention. Is there confusion between two distinct though closely related conceptions, or is there merely an arbitrary use of the two terms as synonyms? For, at first sight, there seems to be no choice but between these alternatives.

Let us begin by taking a mathematician's own view of the distinction between number and quantity, expressed in a perfectly simple and clear manner. The distinction is drawn in the course of a careful explanation of the mathematical theory of measurement, as follows:

'NUMBER and QUANTITY. When the unit is stated the magnitude of the object is precisely determined by its measure in terms of the unit, and this measure is always a number. The "object" may be anything which we can think of as measurable in respect of any property, and the phrase "magnitude of an object" is thus co-extensive in meaning with the word "quantity". The quantity does not change when the unit chosen to measure it changes, and the quantity is not identical with the number expressing it.

'A number can express a quantity only when the unit of measurement is stated or understood. When the unit is stated or implied the number expresses the quantity.'[1]

In this explanatory statement the terms Number, Quantity, Magnitude, and Measure are all involved, and the explanation makes the conceptual distinctions intended in the use of these terms as clear as any one could wish, though perhaps it would

[1] *Theoretical Mechanics*, by A. E. H. Love, M.A., F.R.S. (Univ. Press, Camb.), p. 372.

have been better to have substituted for 'expressing' and 'express', in the first and second paragraphs respectively, the words 'measuring' and 'measure' in accordance with the definition of the measure of a quantity as being always a number. The distinction is again quite clearly marked in the two paragraphs which follow immediately upon those already quoted:

'Mathematical equations, and inequalities, are relations between numbers, expressing that a certain number which has been arrived at in one way is equal to, greater than, or less than, a certain number which has been arrived at in another way.

'Mathematical equations, and inequalities, between numbers expressing quantities are valid expressions of relations between the quantities only if they hold good for all systems of units.'

The contrast between these two paragraphs clearly exhibits the distinction between the science of number and the application of that science to theorems and problems involving the measurement of quantities. In the former the subject of reasoning is number, in the latter it is both number and quantity. From this point of view it is therefore pertinent to inquire why the term 'quantity', which is intended to express a meaning clearly distinct from that of the term 'number', should be used in the science of number without clear indication of the sense in which it is to be taken.

J. B. Stallo, the author of *Concepts of Modern Physics*, makes the following remarks (p. 265 of that work) on this apparent incongruity in the terminology of pure mathematics:

'The error respecting the true nature and function of arithmetic and algebraic quantities has become next to ineradicable by reason of the inveterate use of the word "quantity" for the purpose of designating indiscriminately both extended objects or forms of extension and the abstract numerical units or aggregates by means of which their metrical relations are determined. The effect of this indiscriminate use is another illustration of the well-known fact in the history of cognition that words react powerfully on the thoughts of men, and by this reaction become productive of incalculable error and confusion. It is not to be expected, of course, that mathematicians will cease, at this late day, to speak of arithmetical and algebraic symbols as "quantities"; but there may be some hope for the suggestion that they might return to the old phrase "geometrical (and other) magnitudes". The mischief lies, not so much in the use of a particular word, as in the employment of the same word for the denotation of objects differing from each other *toto genere.*'

This criticism has an obvious bearing on the question: why do algebraists regard a generalized symbol of number as also a symbol of abstract quantity? By implication, if not directly, Stallo condemns the practice as leading to error or confusion, for there is in this respect no difference between a word and a mathematical symbol. I may say, in passing, that I do not very well see how the confusion—if there is any—is to be avoided by the substitution which Stallo recommends; but this is probably because I cannot find any evidence that this custom really does give rise to confusion of thought. The fact is, I believe, that this practice, common to all algebraists, results from a tacit process of thought or judgement which needs only to be made explicit in order to afford a justification or explanation of the habit. A number expresses or measures a quantity only, as Professor Love says, when the unit of measurement is stated or understood; and the quantity, i.e. magnitude of the object, is not identical with the number expressing it because this number changes with any change in the unit of measurement. But this distinction becomes quite indefinite when abstraction is made of any measurable 'object', and we think merely of quantity in the abstract or quantitative relation in general as expressed by number. If it is admissible to regard numerical relation as expressive of quantitative relation in general, then the numerical unit becomes identical with the unit of abstract quantity, and generalized symbols of number can properly be considered as also symbols of abstract quantity.

CHAPTER V

SCOPE AND CHARACTER OF THE ENSUING DISCUSSION

WHAT are we to think of, how shall we characterize, a mental process which might, briefly and in general terms, be indicated thus: Explanation of the derivation, from a primary conception (say that of Quantity, or again, of Space), of another conception, followed by questions such as these: What is, or What is the nature of, this derived conception? Or—say that Abracadabra is the name given to this derived conception—What is the meaning of Abracadabra? The questions imply that the nature of a derived conception is not made manifest in the account of its derivation, that the meaning of a term may be something other than that which we have agreed to assign to it. Yet if a derived conception really is present to the mind, to ask what this conception is can be only an indirect way of asking for an explanation of the process by which we have come to form it, that is, of its derivation; and, if we have given the name Abracadabra to a certain idea, to ask, What is Abracadabra? is to imply that the meaning of this term is not that idea. But select what term you please to characterize a mental process such as this—I can find none more appropriate than 'mystical'—and it will probably seem to become inappropriate when you learn further that a notion which is derived, rather than that from which it derives, is the fundamental notion of the subject of thought which involves them both.

This is not a fancy picture, or a perverse interpretation, but a genuine impression, conveyed in general terms, of an important part of Cayley's presidential address to the British Association in 1883. That Cayley was himself not altogether satisfied with the explanation which he there gave of the doctrine of Imaginaries in mathematics, he made plain by remarking in the course of it that, so far as he knew, the subject had never yet been adequately discussed, and stood in need of a philosophical foundation or justification.

We are supposed, in the calculus, and by means of its symbolism, to develop the 'notion' of imaginary quantity or magnitude. As a counterpart to the notion of imaginary quantity in the calculus, we have, or are supposed to have, in that other great domain of mathematics, geometry, the notion of imaginary *loci* in space—a notion, according to Cayley, as fundamental for geometry as is that of imaginary quantity in analysis. But this is not all. We are supposed also to have an idea or notion of modification of space; that is, a modification or extension of the ordinary conception of space: a more general conception susceptible of specific determinations. According to the modern doctrine, this general conception of externality embraces particular conceptions of space, indicated by the use of such expressions as Euclidean or homaloid space, non-Euclidean or curved spaces of several varieties, and, in each species or variety, spaces of one, two, three, or more dimensions.

Now, that collocations of words such as 'imaginary magnitude', 'imaginary locus', 'homaloid and curved space', do not in my mind evoke modification of my conceptions of magnitude, locality, and space, while on the contrary they are, or seem to be, for modern mathematicians, symbolic and expressive of such modifications—this is a fact which, I must admit, may perfectly well be due merely to inability on my part to follow the conceptual development which is said to issue in these modifications. Upon this point I cannot have anything to say; my part is to criticize, to the best of my ability, the exposition of this development, and leave it to those competent in the matter to judge whether I follow it or not, in so far as it is genuine.

As a result of the general criticism contained in the chapter which follows this, I found myself compelled to assume, as at least probable, that mathematicians do not, in fact, attain to these alleged modifications or extensions of the ordinary ideas of magnitude, locality, and space; and that, where they believe that they do so, that belief results from an illusion of judgement as to the part which symbolism of any kind plays in the development of a process of reasoning; in other words, is the result of a tendency to mysticism of which they are unconscious, or not sufficiently conscious. Now it is in the nature of the case that, while we can recognize an illusion of this kind in our own mind, we can never afford a direct proof of its existence in any other. But, having established by general and indirect evidence a reasonable prima facie case for the suspicion, we shall greatly add to the weight of this evidence if we can show that throughout the alleged derivation of these notions, as set forth in the received exposition, and from the very outset of it, the mind of the learner is subjected (quite innocently, of course) to a mystical bias which gradually and insensibly modifies the natural course of plain unsophisticated reasoning, and by easy transitions prepares him to admit, without too violent a shock to his sense of the rational, the reality of these so-called fundamental notions.

The whole subject divides itself naturally into two parts which have no necessary logical connexion : (1) the doctrine of imaginary magnitude in analysis and of imaginary *loci* in space ; (2) the doctrine of a generalized conception of space, or of the conceiving of different kinds of space together with the different geometrical systems which they involve. In this part we are concerned only with the former of these two doctrines.

CHAPTER VI

THE DOCTRINE OF MATHEMATICAL IMAGINARIES

Cayley's explanation of the Doctrine.—His plea for a philosophical discussion of it.—What is meant by the 'meaning' of a notion ?—Cayley's metaphysical outlook upon Geometry.—Mr. A. N. Whitehead's explanation of the Doctrine.—This explanation appears to be founded upon a theory of signs not reconcilable with our mental processes.—It does not differ essentially from the explanation given by Boole in *The Laws of Thought*.—Boole's explanation, however, is rather a begging of the question than a solution of the enigma.

I REFERRED in the last chapter to Cayley's invitation to thinkers to put the doctrine of Imaginaries upon a sound philosophical basis. I call it an invitation rather than a challenge, as it has sometimes been called, because it proceeded not from a disbeliever in the doctrine, but, on the contrary, from one who, while profoundly convinced that it possesses genuine significance, was unwilling to hide, under a specious dialectic, the incongruities of thought apparently involved in it. In a discussion of the origin of these incongruities, in an attempt to discover whether they are real or only apparent, whether they are the embodiment of some real illusion of judgement; or, possibly, issue merely from an inadequacy of the ordinary forms of language to express definite but unfamiliar relations of thought—it will be of manifest advantage to start with the simple and clear exposition, given in the words of the great mathematician himself, of the way in which these incongruities presented themselves to him. I shall therefore begin by quoting, from Cayley's Presidential Address, the passages in one of which the invitation referred to is contained:

'In Arithmetic and Algebra, or say in analysis, the numbers or magnitudes which we represent by symbols are in the first instance ordinary (that is, positive) numbers or magnitudes. We have also in analysis and analytical geometry *negative* magnitudes; there has been in regard to this plenty of philosophical discussion, and I might refer to Kant's paper, *Ueber die negativen Grössen in der Weltweisheit* (1763), but the notion of a negative magnitude has become quite a familiar one, and has extended itself into common phraseology. . . . But it is far otherwise with the notion which is really the fundamental one (and I cannot too strongly emphasize the assertion) underlying and pervading the whole of modern analysis and geometry, that of imaginary magnitude in analysis and of imaginary space (or space as a *locus in quo* of imaginary points and figures) in geometry: I use in each case the word imaginary as including real. This has not been, so far as I am aware, a subject of philosophical discussion or inquiry . . . considering the prominent position which the notion occupies—say even that the conclusion were that the notion belongs to mere technical mathematics or has reference to nonentities in regard to which no science is possible, still it seems to me that (as a subject of philosophical discussion) the notion ought not to be thus ignored; it should at least be shown that there is a right to ignore it.'[1]

After a digression on the subject of non-Euclidean geometry, Cayley proceeds:

'Coming now to the fundamental notion already referred to, that of imaginary magnitude in analysis and imaginary space in geometry, I connect this with two great discoveries in mathematics made in the first half of the seventeenth century, Harriott's representation of an equation in the form $f(x) = 0$, and the consequent notion of the roots of an equation as derived from the linear factors of $f(x)$, . . . and Descartes method of co-ordinates, as given in the *Géométrie*. . . .'

'Taking the coefficients of an equation to be real magnitudes, it at once follows from Harriott's form of an equation that an equation of the order n ought to have n roots. But it is by no means true that there are always n real roots. In particular, an equation of the second order, or quadric equation, may have no real roots; but if we assume the existence of a root i of the quadric equation $x^2+1 = 0$, then the other root is $-i$; and it is easily seen that every quadric equation (with real coefficients as before) has two roots, $a \pm bi$, where a and b are real magnitudes. We are thus led to the conception of an imaginary magnitude, $a+bi$, where a and b are real magnitudes, each susceptible of any positive or negative value, zero included. The general theory is that, taking the coefficients of the equation to be imaginary magnitudes, then an equation of the order n has always n roots, each of them an imaginary magnitude. . . . The idea is that of considering, in place of real magnitudes, these imaginary or complex magnitudes $a+bi$.

'In the Cartesian geometry a curve is determined by means of the equation existing between the co-ordinates (x, y) of any point thereof. In the case of a right line this equation is linear; in the case of a circle, or more generally of a conic, the equation

[1] Cayley, *Collected Mathematical Works*, vol. xi, p. 434.

is of the second order; and generally, when the equation is of the order n, the curve which it represents is said to be a curve of the order n. In the case of two given curves, there are thus two equations satisfied by the co-ordinates (x, y) of the several points of intersection, and these give rise to an equation of a certain order for the co-ordinate x or y of a point of intersection. In the case of a straight line and a circle, this is a quadric equation; it has two roots, real or imaginary. There are thus two values, say of x, and to each of these corresponds a single value of y. There are therefore two points of intersection—viz. a straight line and a circle intersect *always* in two points, real or imaginary. It is in this way that we are led analytically to the notion of imaginary points in geometry. The conclusion as to the two points of intersection cannot be contradicted by experience: take a sheet of paper and draw on it the straight line and circle, and try. But you might say, or at least be strongly tempted to say, that it is meaningless. The question of course arises, What is the meaning of an imaginary point? and further, in what manner can the notion be arrived at geometrically?'[1]

Cayley, as we see, does not here put the question, What is the meaning of imaginary magnitude?—a question which is of course quite unnecessary if in his exposition he has shown us how we come to form, or whence we derive, a certain notion called that of imaginary magnitude. But, if he has done this, it would seem that we already have (so far as imaginary magnitude is concerned) that which Cayley avers that we lack, viz. a philosophical explanation of the notion; unless, indeed, a philosophical inquiry about a notion is something quite different from an inquiry into its origin and derivation. The peculiarity of the process of thought is more obvious in relation to imaginary *loci*, because there Cayley does overtly ask the question, What is the meaning of an imaginary point? after having carefully explained to us how mathematicians are led analytically, i.e. by means of algebraic symbolism, to the notion of imaginary points. If the explanation is a real one, what more can philosophy do for us in the matter? If Cayley or any one else has in reality been led, no matter by what particular process of thought, conducted by what system of symbolic expression soever, to form a notion to which he gives this name 'imaginary points', then the meaning of, the thought expressed by, these words 'imaginary points', *is* that notion which he has thus formed. How, then, does the

[1] *Op. cit.*, pp. 437, 438.

'question of course arise, What is the meaning of an imaginary point?' But if, on the other hand, neither Cayley nor any one else has in reality been led to form any modification or extension of the ordinary conception of geometrical locus; and if what has happened is that, through some illusion concerning the instrumentality of a symbolic system (whether that of ordinary language or any other), we have been led merely to *believe* that we have attained to some such modification or extension; then, indeed, nothing would be less surprising than that this question, What is the meaning of an imaginary point? should thus present itself.

Cayley does not answer this question; and yet he does answer, or seems to answer, the further question, How is this notion arrived at geometrically?

'There is a well-known construction in perspective for drawing lines through the intersection of two lines, which are so nearly parallel as not to meet within the limits of the sheet of paper. You have two given lines which do not meet, and you draw a third line, which, when the lines are all of them produced, is found to pass through the intersection of the given lines. If instead of lines we have two circular arcs not meeting each other, then we can, by means of these arcs, construct a line; and if on completing the circles it is found that the circles intersect each other in two real points, then it will be found that the line passes through these two points: if the circles appear not to intersect, then the line will appear not to intersect either of the circles.[1] But the geometrical construction being in each case the same, we say that in the second case also the line passes through the two intersections of the circles.'

He then adds:

'Of course it may be said . . . that the conclusion is a very natural one, provided we assume the existence of imaginary points; and that, this assumption not being made, then, if the circles do not intersect, it is meaningless to assert that the line passes through their points of intersection!'

But he leaves the matter there, concluding with the remark:

'As a matter of fact, we do consider in plane geometry imaginary

[1] Is there not here just a touch of the sophistical or ambiguous in the way the two words 'found' and 'appear' are used? Compare with the sentence (in the previous excerpt): 'The conclusion as to the two points of intersection cannot be contradicted by experience'—in which the use of the term 'experience' is in the highest degree ambiguous.

points introduced into the theory analytically or geometrically as above.'

A plainer statement of the matter of fact there could not well be. No one, however, doubts the fact; what we want is an explanation of it. Does Cayley give it to us? It is not easy to say what he himself thought about the matter. Yet, *apparently*, the explanation given is not only simple, but adequate:

'But the geometrical construction being in each case the same, we say that in the second case also the line passes through the two intersections of the circles.'

That is to say, the geometer perceives, in the construction common to these two opposed cases, a certain analogy; and this analogy is paradoxically, or by a violent metaphor, expressed in the statement that the line always passes through the intersections, real or imaginary, of the two circles. But then, so far as the expression 'imaginary points' alone is concerned, this is the philosophy of the matter. We require nothing more, save to recollect that we have expressed a real analogy by means of a verbal paradox, and that we must be careful, especially in the development of such an unusual mode of expression, not to lapse into the mystical by subsequently trying to read these expressions as if they were literal.

This, however, does not appear to be the mathematician's point of view, so far at least as Cayley is here representative of it, for he remains in search of a philosophy. But what, then, is this point of view?

If we take Cayley's explanation as a whole, with regard both to imaginary magnitude and imaginary *loci*, the point of view which it suggests is something of this kind: We are led, in the way described, to these notions; and when we ask, What is imaginary magnitude? What are imaginary points?—these questions arise from our wish to know, if it is knowable, to what realities these notions correspond, or whether they do not correspond to any reality at all, have 'reference to nonentities in regard to which no science is possible.' Such an attitude of mind, comprehensible enough were we considering the (current) fundamental notions of the physical, or of the biological sciences, seems to me here to be purely mystical, and possibly to be accounted for by the survival, in the adult mind and in attenuated form, of what, in the child's mind, is a strong disposition indiscriminately to believe in the existence of things and ideas on the bare evidence of names, and of relations of ideas on the bare evidence of phrases which do not involve obvious contradictions.

But, again, it is not at all uncommon even for careful thinkers to speak of the 'meaning' of an idea, notion, or conception—a mode of expression which does not seem to me happy, but which is perhaps unavoidable in a transitional state of philosophical terminology. I mention this in relation to Cayley's question because it suggests a possible interpretation of it, though it seems to me evident that it can afford no real clue to his mental attitude. To ask what the meaning is of a notion is, I suppose, to inquire of what value or import it is in the subject or scheme of thought in which it is involved, in what relations it stands to other notions in that scheme. There is nothing at all incongruous in such a question, and the answer to it must be found in an adequate analysis of the conceptual development of the particular subject. Such an analysis, however, is just what Cayley appears to give us: he traces the development of these notions from antecedent notions. We cannot, then, suppose that by the question with which he terminates this analysis he intended to ask in what relations these notions stand to the others.

I will conclude this survey of Cayley's outlook upon the doctrine of imaginaries by drawing attention to an earlier passage in his Presidential Address not directly connected with this doctrine, but interesting in the clear indication it contains that for him the approach to a true understanding of the fundamental notions of mathematics lay through the gateway of metaphysical speculation. In this part of his address Cayley considers at some length J. S. Mill's views as to the nature of mathematical, especially geometrical, truths, and he sums up his own conclusions thus:—

'I think it may be at once conceded that the truths of geometry are truths precisely because they relate to and express the properties of what Mill calls "purely imaginary objects";[1] that these objects do not exist in Mill's sense, that they do not

[1] Mill's 'purely imaginary objects' have no connexion whatever with the imaginaries we have been discussing; they are simply the points, lines, figures, &c., of Geometry, as defined e.g. in Euclid's *Elements*.

exist in nature, may also be granted; that they are "not even possible", if this means not possible in an existing nature, may also be granted. That we cannot "conceive" them depends on the meaning which we attach to the word conceive. I would myself say that the purely imaginary objects are the only realities, the ὄντως ὄντα, in regard to which the corresponding physical objects are as the shadows in the cave; and it is only by means of them that we are able to deny the existence of a corresponding physical object; if there is no conception of straightness, then it is meaningless to deny the existence of a perfectly straight line.'[1]

I do not know how far these views are held by philosophical mathematicians; but it is not difficult to see how one may be brought round to Cayley's opinion. The origin of it evidently lies in what must seem to be an impossibility of deriving these notions (the purely imaginary objects) from experience. If we find it impossible thus to account for their presence in our minds, we may very easily be driven or tempted to conclude that they are independent of our geometrical experience, do not result from it, but are *a priori conditions* of that experience or of the judgements involved in it—to use Kantian phraseology—or, as Cayley implies, it is only his conception of straightness which, contrasted with his experience of linear shape, enables him to deny the existence of a perfectly straight line; in other words, this conception is an *a priori* condition of his judgement of linear shape.

But does it follow that because we have had no experience of absolute straightness the conception of it cannot have originated in experience itself. Not only does this not follow, but it is, I believe, quite possible to show that the nature of our experience of shape is such that this and the other notions—the purely imaginary objects—must necessarily arise in that experience. The discussion of this, however, is relevant to the third part of this work. I leave it for future consideration, and return to what more immediately concerns us.

The question which troubled Cayley had, now in a desultory, and now in a persistent way, occupied the attention of mathematicians since the time when first symbols of so-called imaginary quantity introduced themselves into algebraic symbolism, as it were, in despite of the symbolists and of common sense. The

[1] *Op. cit.*, p. 433.

enigma was a perpetual challenge to the philosophically-minded in the mathematical world. In the early years of the nineteenth century, and more or less throughout the first half of it, a great deal of thought was expended over this riddle. But, quite obviously, these labours had failed, in Cayley's opinion, to dispose of the difficulty in a satisfactory way. And not only in Cayley's opinion. Mr. Bertrand Russell, in his work on *The Foundations of Geometry*, published in 1897, fourteen years after the date of Cayley's Address, referring to Cayley's invitation or challenge, says, in the course of an interesting discussion on the philosophy of geometrical imaginaries, that he is 'unacquainted with any satisfactory philosophy of imaginaries in pure algebra'.[1] On the other hand, Mr. A. N. Whitehead appears to find,[2] in the work of the mathematicians of the first half of the nineteenth century, a satisfactory philosophy of the subject; but I am not quite clear, from the form which his explanation takes, whether he puts it forward as one to be accepted, or as one already recognized and accepted by mathematicians generally. However, this point is of little importance as compared with that of the explanation itself, which is especially interesting from its author's clear recognition that any theory of the matter, to be of some value, must be founded upon a definite view of the *rationale* of symbolism in the process of thought.

Such a view, unfortunately too briefly and didactically expressed to lend itself very easily to effective comment, Mr. Whitehead gives us in the introductory chapter of the work already mentioned. We must look upon it, such as it is, as the key to the explanation which he subsequently gives of the doctrine of imaginaries. Its essential traits are, I think, clearly enough indicated in the following passages:

'A substitutive sign is such that in thought it takes the place of that for which it is substituted. A counter in a game may be such a sign: at the end of the game the counters lost or won may be interpreted in the form of money, but till then it may be convenient for attention to be concentrated on the counters and not on their signification. The signs of a Mathematical Calculus are substitutive signs.'[3] . . .

[1] *An Essay on the Foundations of Geometry.* Cambridge: University Press, 1897, p. 43. But see also footnote on p. 83 of the present work.

[2] *Universal Algebra*, by A. N. Whitehead, Sc.D., F.R.S.

[3] Mr. Whitehead then quotes Dr. Stout's dictum: 'A word is an instru-

'The use of substitutive signs in reasoning is to economize thought.'

'Definition of a Calculus. In order that reasoning may be conducted by means of substitutive signs, it is necessary that rules be given for the manipulation of the signs. The rules should be such that the final state of the signs after a series of operations according to rule denotes, when the signs are interpreted in terms of the things for which they are substituted, a proposition true for the things represented by the signs. The art of the manipulation of substitutive signs according to fixed rules, and the deduction therefrom of true propositions, is a calculus. . . .

'When once the rules for the manipulation of the signs of a calculus are known, the art of their practical manipulation can be studied apart from any attention to the meaning to be assigned to the signs. It is obvious that we can take any marks we like and manipulate them according to any rule we choose to assign. It is also equally obvious that in general such occupations must be frivolous. They possess a serious scientific value where there is a similarity of type of the signs and of the rules of manipulation to those of some calculus in which the marks used are substitutive signs for things and relations of things. The comparative study of the various forms produced by variation of rules throws light on the principles of the calculus. Furthermore, the knowledge thus gained gives facility in the invention of some significant calculus designed to facilitate reasoning with respect to some given subject.

'It enters therefore into the definition of a calculus properly so-called that the marks used in it are substitutive signs. But when a set of marks and the rules for their arrangements and re-arrangements are analogous to those of a significant calculus, so that the study of the allowable forms of their arrangement throws light on that of the calculus—or when the marks and their rules of arrangement are such as to appear likely to receive an interpretation as substitutive signs or facilitate the invention of a true calculus, then the art of arranging such marks may be called—by an extension of the term—an uninterpreted calculus. The study of such a calculus is of scientific value. The marks used in it will be called signs or symbols as are those of a true calculus, thus tacitly suggesting that there is some unknown interpretation which could be given to the calculus.'[1]

There is in this doctrine so much which seems to me to require explanation that I may be pardoned the playful exaggeration of saying that it suggests analogy with an uninterpreted calculus

ment for thinking of the meaning which it expresses; a substitute sign is a means of not thinking of the meaning which it symbolizes.' See *passim* my remarks on this (p. 37).

[1] *Universal Algebra*, pp. 3–5.

and the unknown interpretation which might be given to it. As a theory of the connexion or relations between rational thought, whether mathematical or other, and its symbolization in any form, it is not convincing; but then, it is true, in such a theory there would not appear to be room for the invention of a non-interpreted symbolic system, an uninterpreted calculus. Whatever it be that the doctrine seeks to set forth or explain, it leaves an abyss, unplumbed and unbridged, between a calculus 'properly so called' and something which, 'by an extension of the term,' may be called an uninterpreted calculus.

But what is the real gist of the argument? The definition of a substitutive sign is, to begin with, inconveniently vague if not unmeaning. A substitutive sign is such that in thought it takes the place of that for which it is substituted. In other words, a substitutive sign is such that in thought it takes the place of that whose place it takes, or is made to take. That is not sufficiently informing; it is true of anything you please and descriptive of nothing in particular; we know neither what 'that' is whose place is taken, nor in which of half a dozen possible senses we are to take the word 'substitution' or the phrase 'to take the place of'. And this is not hypercriticism: it would have been so very easy for Mr. Whitehead to say: a substitutive sign is a sign which we intend to take the place of (i.e. to be equivalent in meaning with) another sign or combination of signs, that as he does not say this his meaning is doubtful; and the doubt is not removed by the illustration of counters in a game. On the other hand it is not impossible to guess why we are given the above elusive definition. What is commonly understood by a substitutive sign is a sign substituted for words, but words have meanings, and this is awkward if substitutive signs may have none.

In general, as Mr. Whitehead observes, to take a set of marks and manipulate them according to any rule we may please to lay down must be a frivolous occupation. That would appear to be the case where either the marks, or the manipulations according to rule, do not symbolize anything at all. But then, if the marks are not symbolic they are not marks but things, and where the rule of manipulation is not symbolic it is at once arbitrary and meaningless—is not, in any sense relevant to symbols, a rule of manipulation.

But such occupations, it is said, possess serious scientific value when there is a similarity of type of the signs and of the rules of manipulation to those of some calculus in which the marks used are substitutive signs for things and relations of things. I do not doubt it ; only there is one rather important point which is ignored in this explanation. How or why does one set of signs and of the rules for their manipulation happen to be similar to another set unless there is analogy of conception and of thought-process seeking and finding expression in this similarity of type of sign and of rule ; in other words, unless the later calculus is significant ? It is also pertinent to inquire how, in a non-significant calculus, one form of arrangement of the signs can be 'allowable' and another not. Mr. Whitehead speaks of symbolic systems in turns of phrase which recall those of a geologist comparing the fossil remains of different strata, or of a zoologist describing the fauna of different continents. But symbolic systems are not found, they are invented ; purpose presides at their invention, and that purpose is the systematic expression of a process of reasoning.

In the 'Definition of a Calculus' there is one other statement, the exact import of which we ought to be quite clear about, for it is closely connected with the riddle of Imaginaries in algebra. I mean the statement that 'the rules should be such that the final state of the signs after a series of operations according to rule denotes, when the signs are interpreted in terms of the things for which they are substituted, a proposition true for the things represented by the signs.' 'The *final* state of the signs' implies, and is here, I believe, intended to imply, 'not necessarily the intermediate states.' The implication is, of course, not that the intermediate states may denote false propositions, but that they need not be interpretable.

This doctrine derives from Boole, whether originally or not I do not know. But do the arguments by which Boole supports this doctrine amount to anything more than an assertion, or a plea, that what appear to be incongruities in a process of thought are in reality not such ? If we accept Boole's view it practically determines our attitude in the question of these imaginaries, and therefore I had better explain here why I cannot agree with it. This is what Boole says :

'The conditions of valid reasoning, by the aid of symbols, are—

'1st. That a fixed interpretation be assigned to the symbols employed in the expression of the data; and that the laws of the combination of those symbols be correctly determined from that interpretation.

'2nd. That the formal processes of solution or demonstration be conducted throughout in obedience to all the laws determined as above, without regard to the question of the interpretability of the particular results obtained.

'3rd. That the final result be interpretable in form, and that it be actually interpreted in accordance with that system of interpretation which has been employed in the expression of the data.'

The first and third of these conditions seem to be unimpeachable, but the second Boole himself felt would be accepted with difficulty. He comments on it as follows:

'It is . . . in connexion with the second of the above general principles or conditions, that the greatest difficulty is likely to be felt, and upon this point a few additional words are necessary. . . . The principle in question . . . is derived, like the knowledge of the other laws of the mind, from the clear manifestation of the general principle in the particular instance. . . . The employment of the uninterpretable symbol $\sqrt{-1}$, in the intermediate processes of trigonometry, furnishes an illustration of what has been said. I apprehend that there is no mode of explaining that application which does not covertly assume the very principle in question. But that principle, though not, as I conceive, warranted by formal reasoning based upon other grounds, seems to deserve a place among those axiomatic truths which constitute the foundation of the possibility of general knowledge, and which may properly be regarded as in some sense expressions of the mind's own laws and constitution.'[1]

But if, as Boole says, the knowledge of this law of the mind is derived, like the knowledge of the other laws of the mind, from the clear manifestation of the general principle in the particular instance, how is it that great difficulty is felt with regard to it and not with regard to the others? Simply because there is no clear manifestation of a general principle in the particular instance. It is clear that the distinction between an intermediate result and the final result in a process of reasoning is not an essential distinction so far as the process itself is concerned: the process may be continued, and what was a final result may become an intermediate one. If it were a matter of experience

[1] Boole, *Laws of Thought*, pp. 68, 69.

that intermediate results in a process of reasoning were always, or even usually, uninterpretable, then we should feel no difficulty —we should recognize the general principle in the particular instance. For my part I recognize in this case just the contrary: that the particular instance manifests an unexplained departure from general principle.

We can now go on to consider Mr. Whitehead's elucidation of the enigma involved in the use of algebraic imaginaries:

'The logical difficulty involved in the use of a calculus only partially interpretable can now be explained. The discussion of this great problem in its application to the special case of $(-1)^{\frac{1}{2}}$ engaged the attention of the leading mathematicians of the first half of this century, and led to the development on the one hand of the Theory of Functions of a Complex Variable, and on the other of the science here called Universal Algebra.

'The difficulty is this: the symbol $(-1)^{\frac{1}{2}}$ is absolutely without meaning when it is endeavoured to interpret it as a number; but algebraic transformations which involve the use of complex quantities of the form $a+bi$, where a and b are numbers, and i stands for the above symbol, yield propositions which do relate purely to number. As a matter of fact the propositions thus discovered were found to be true propositions. The method therefore was trusted, before any explanation was forthcoming why algebraic reasoning, which had no intelligible interpretation in arithmetic should give true arithmetical results.

'The difficulty was solved by observing that Algebra does not depend on Arithmetic for the validity of its laws of transformation. If there were such a dependence, it is obvious that as soon as algebraic expressions are arithmetically unintelligible all laws respecting them must lose their validity. But the laws of Algebra, though suggested by Arithmetic, do not depend on it. They depend entirely on the convention by which it is stated that certain modes of grouping the symbols are to be considered as identical. This assigns certain properties to the marks which form the symbols of Algebra. The laws regulating the manipulation of the algebraic symbols are identical with those of Arithmetic. It follows that no algebraic theorem can ever contradict any result which could be arrived at by arithmetic, for the reasoning in both cases merely applies the same general laws to different classes of things. If an algebraic theorem is interpretable in arithmetic, the corresponding arithmetical theorem is therefore true. In short, when once Algebra is conceived as an independent science dealing with the relations of certain marks conditioned by the observance of certain conventional laws, the difficulty vanishes. If the laws be identical, the theorems of the one science can only give results conditioned by the laws which also hold good for the other science; and therefore these results, when interpretable, are true.'[1]

While admitting that this solution is consistent with the views previously expressed by Mr. Whitehead on the nature and functions of a mathematical calculus, we may ask ourselves whether those views are not themselves a generalization of that here adopted as a solution in the case of ordinary algebra; whether, in short, we have not here a solution of the difficulty after the manner of Boole. If we are unable to agree with Mr. Whitehead's general conception of the functions of a symbolic system in the conduct of a process of reasoning—and I have given my own reasons, be they good or bad, for dissent—then naturally enough we shall see, in this solution of the enigma, merely a restatement of it in a thinly disguised form. The difficulty was solved by observing something which had previously been overlooked, viz. that Algebra does not depend on Arithmetic for the validity of its laws of transformation. Is this anything more than a restatement of the observed fact that algebraists in generalizing the symbolism and operations of arithmetic, had been led to invent symbolic forms not arithmetically interpretable? The ordinary explanation of this fact, that which Cayley gave but was not satisfied with, and which is still given in authoritative textbooks of algebra, is that the development of this generalized symbolism itself leads to a non-arithmetical notion of quantity or magnitude. This, at least, is an attempt, though a mystical attempt, at explanation.

Algebra is a symbolic system which has been in course of gradual development for centuries. The term itself is a Western adaptation of an Arabic expression which, judging by the etymology given of it, might just as well have been translated 'arithmetic' as corrupted into 'algebra'. The Moslem and other *savants* from whom the system was borrowed and introduced into Europe soon after 1200, were themselves little more than depositaries or inheritors of neo-Greek and possibly of Hindoo learning. The treatise of Diophantos may be and has been indifferently called an arithmetical and an algebraic treatise. The fact (if it is a fact) that algebraic symbolism had in the

[1] *Op. cit.*, pp. 10, 11.

course of its development become independent or partially independent of arithmetical notions is not explained by observing it; it can only be explained either by showing what notions the symbolism had become dependent on, or by showing that the observed independence is not real, but only apparent. We come back upon the undisguised fundamental difficulty: What is the meaning of 'imaginary number', 'imaginary quantity', '$\sqrt{-1}$'?

These symbols, it is agreed, are not symbols of notions which enter into Arithmetic. They are, according to some mathematicians, symbols of notions which do enter into Algebra—apparently without first entering into our minds. According to Mr. Whitehead they are symbols uninterpretable not only in Arithmetic, but in Algebra itself; they happen to receive an interpretation in Geometry, but the laws of Algebra do not depend for their validity on these symbols being interpretable in that science or in any other. Upon what, then, do these laws depend? 'They depend entirely on the convention by which it is stated that certain modes of grouping the symbols are to be considered as identical.' This statement, apparently simple, I find to be ambiguous with regard (1) to the definition given of the convention, (2) to the use made of the word 'depend'. The ambiguity of the definition lies in this: that it is impossible to say whether it is or is not intentionally devised so as to be verbally congruous with the enigma of a calculus some of whose symbols are uninterpretable. It is quite easy to mark the ambiguity by inquiring whether the given definition of the convention is equivalent to this: that certain non-identical modes of grouping the symbols are intended to symbolize, and are accepted as symbolizing, the same or equivalent notions. Next, as regards the use of the word 'depend'. If we pause to consider what is the nature of these 'laws' of algebraic transformation, we see that they are simply statements of the conventions which sanction the transformations: they are 'laws' just in so far as they are adopted conventions. The validity of the laws is just the validity of the conventions; and to make the validity of the laws depend upon the conventions has thus no clear meaning. Yet this seems to be what Mr. Whitehead does. The validity of these conventions or laws depends, I should say, upon their adaptability to the end in view, which is the unambiguous symbolization of a process of reasoning.

I do not wish to multiply objections; but I must add, that when Mr. Whitehead tells us that: 'In short, when once Algebra is conceived as an independent science dealing with the relations of certain marks conditioned by the observance of certain conventional laws, the difficulty vanishes,' he sees neither that the difficulty is precisely that of so conceiving Algebra, or any other symbolic system, nor that to say that it is to be thus conceived is hardly reconcilable with his own previous statement (concerning the impossibility of contradiction between Arithmetic and Algebra) that 'the reasoning in both cases merely applies the same general laws to different classes of things.' For evidently if there *is* a class of 'things' about which we reason by means of Algebra, then we do *not* conceive Algebra as merely dealing with the relations of certain marks conditioned by the observance of certain conventional laws, but as a symbolic system dealing, by means of these marks and laws, with the 'class of things' which is the subject of thought.

I will conclude this criticism by a last quotation in justification of a statement made on a previous page:

'It will be observed that the explanation of the legitimacy of the use of a partially interpretable calculus does not depend upon the fact that in another field of thought the calculus is entirely interpretable. The discovery of an interpretation undoubtedly gave the clue by means of which the true solution was arrived at. For the fact that the processes of the calculus were interpretable in a science so independent of Arithmetic as Geometry at once showed that the laws of the calculus might have been defined in reference to geometrical processes. But it was a paradox to assert that a science like Algebra, which had been studied for centuries without reference to Geometry, was after all dependent upon Geometry for its first principles. The step to the true explanation was then easily taken.'[1]

This paragraph makes quite clear what may have seemed hitherto somewhat obscure—I refer to my statement above that for Mr. Whitehead such symbolic expressions as 'imaginary quantity', '$\sqrt{-1}$,' are uninterpretable not only in Arithmetic but also in Algebra. This, of course, is quite consistent if Algebra is a calculus only partially interpretable in its own field of thought (whatever this may be); it is also quite consistent with Algebra

[1] *Op. cit.*, p. 11.

being a calculus entirely interpretable in another field of thought, viz. Geometry.

Mr. Whitehead's eminence as a mathematician, especially as a philosophical mathematician, made it very necessary to examine his views on this particular question with the closest attention ; and the necessity appeared to me the more pressing because these views have evidently to some extent—to how great an extent I do not know—imposed themselves on mathematicians.[1] This influence is very marked in the article ' Algebra ' of vol. xxv (one of the new volumes) of the *Encyclopaedia Britannica*. Indeed, the writer of that article explicitly acknowledges his indebtedness to Mr. Whitehead's work. How great that indebtedness is with regard to the subject we have been discussing will be gathered from the following extract (p. 274) :

' The progress of analytical geometry led to a geometrical interpretation both of negative and also of imaginary quantities ; and when a " meaning ", or, more properly, an interpretation, had thus been found for the symbols in question, a reconsideration of the old algebraic problem became inevitable, and the true solution, now so obvious, was eventually obtained. It was at last realized that the laws of algebra do not depend for their validity upon any particular interpretation, whether arithmetical, geometrical, or other ; the only question is whether these laws do or do not involve any logical contradiction. When this fundamental truth had been fully grasped, mathematicians began to inquire whether algebras might not be discovered which obeyed laws different from those obtained by the generalization of arithmetic.'

Although differing somewhat in the form of expression, the views here expressed are in substance Mr. Whitehead's views. The writer of the article, it is true, makes the validity of the laws of algebra depend upon whether they do or do not involve logical contradiction, which is not quite what Mr. Whitehead says. But the difference is rather apparent than real : we very naturally

[1] To a greater extent, it seems, than I suspected when the above sentence was penned. According to Mr. Bertrand Russell (see his *Principles of Mathematics*, Cambridge, 1903, vol. i, p. 376), the theory of imaginaries, formerly considered a very important branch of mathematical philosophy, has lost its philosophical importance by ceasing to be controversial. Mr. Russell and Mr. Whitehead are, I believe, the leading exponents in this country of the latest school of thought on the philosophy of mathematics.

ask how this test of validity is to be applied to a symbolic system unless this system is prompted by, and is therefore expressive of, a systematic process of thought—in other words, unless the system is interpreted? Absence of logical contradiction is certainly a test of valid reasoning; but how either logical contradiction, or logical consistency, is to manifest itself under the supposed conditions is not explained. Obviously, however, no such explanation would be needed by any one who conceived algebra 'as an independent science dealing with the relations of certain marks conditioned by the observance of certain conventional laws': the argument of the two writers is at bottom the same argument.

CHAPTER VII

THE CONCEPTIONS AND SYMBOLISM OF ELEMENTARY ALGEBRA

Current view of the derivation of the conception of Algebraic Quantity.—Real nature of this notion.—The fundamental Laws of Algebraic Expression (Commutation, Association, Distribution) are conventions founded on the symbolic expression of this notion.—To derive this notion from the Law of Association is to put the cart before the horse.—Meaning of the symbol = in relation to the so-called test of inequality.—Algebraic Multiplication and Division.—'Positive' and 'Negative' multiplication and division meaningless when taken literally.—A distorted view of these operations must lead to a distorted view of the operations of Involution and Evolution.

THE difficulty or incomprehensibility involved in the doctrine of imaginaries did not present itself to Cayley, as we saw in the last chapter, in quite the same way as it is presented by Mr. Whitehead. For Cayley, it seems that the difficulty or enigma arises from conflict between a belief, imposed upon us by a logical process of thought, and the direct evidence of consciousness: the belief being that the development of algebraic symbolism leads to a new development of the conception of quantity, while the evidence directly afforded by the consciousness of our own thoughts is that the conception of quantity really remains unmodified. Mr. Whitehead, on the other hand, says nothing about the development of algebraic symbolism leading to a modification or extension of the conception of quantity. In his view, if I have understood it, the difficulty lay in this: that the development of what, in its inception, was a generalized arithmetical symbolism, led to symbolic expressions which have no arithmetical meaning. Which of these two views most correctly represents the difficulty as it was felt by mathematicians in general, I do not know; probably some saw it as Cayley did, others as Mr. Whitehead does. What I propose to do in this chapter and the next is to show, by carefully analyzing the received exposition of the process of conception and symbolization in elementary algebra, that the belief that we are led by this process to a new notion of quantity, or, again, the supposition that the so-called symbols of imaginary quantity ought logically to be arithmetically interpretable, are the result partly of a mystical outlook upon this process, partly of an inconsistency in the symbolic development itself; and that if we admit this inconsistency into the system, clearly recognizing it as such, and also reject whatever is mystical in the exposition, we get rid of the conditions under which enigmatical questions suggest themselves.

With this object in view I believe I cannot do better than to take as a text Professor Chrystal's standard work on Algebra, or the same author's smaller and more elementary work, the *Introduction to Algebra*.[1] As we shall be occupied with no more than the most elementary conceptions and symbolic expressions—and with these only in so far as they lead to the doctrine of Imaginary Quantity—and as these conceptions and expressions are very fully dealt with in the *Introduction*, I shall take this as a text. There are no doubt many other authoritative English textbooks of elementary algebra; but, apart from the acknowledged position of the author of the text chosen as a teacher, it is a manifest advantage to avoid scattering references over a wide field, and to trace the filiation of thought as it presents itself to, and is expressed by, a single mind. I need scarcely add that any criticisms which I may have to make are not to be understood as levelled especially at Professor Chrystal, but at the received exposition of the subject as embodied in a standard work.

One of the first ideas to which the beginner in Algebra is introduced, after the employment of the letters of the alphabet as general symbols for number has been explained, and the conventional signs for the ordinary operations of arithmetic have been exhibited, is that which is termed 'algebraic quantity'. The explanation of this idea is given on p. 18 of the *Introduction*, and is there prefaced by the remark that 'The Law of Association for an algebraic sum . . . leads us to another important idea, viz. the notion of Algebraic Quantity as distinguished from what may be called mere Arithmetical Quantity'. I draw attention to this statement because, with many others of a like kind, it seems to indicate a belief or supposition that conceptual develop-

[1] *Introduction to Algebra*, by G. Chrystal, M.A., LL.D., 2nd edition, 1900.

ment follows upon the development of a symbolic system, rather than that the development of a symbolic system follows upon and is the expression of a development of conception. We would think it a very odd statement were any one to assert that the laws of addition and subtraction in arithmetical symbolism lead us to the notion of number. The above statement is no less odd to any one who keeps clearly in mind the distinction between an idea and the name which is by convention assigned to it. The notion of algebraic summation, as distinct from ordinary or arithmetical addition and subtraction, is really nothing other than the notion of summation of algebraic quantities, whether we use this term for the constituents of the sum, or some other equivalent to it, such as oppositely qualified, or positive and negative, quantities. But the notion itself of quantity as positive or negative is a generalization of, or recognition of a common feature in, such cases as credits and debits, forces acting in opposition to one another, excess over and defect from a mean, &c. It is of course obvious that it was the employment of symbols of quantity and of signs of addition and subtraction which suggested the artifice of using the latter as qualifying signs, that is, of symbolizing this generalized notion of opposition in quantity by the conjunct use of the symbol of addition, or of subtraction, with a symbol of quantity, to denote respectively a positive or a negative quantity. This, however, is very different from saying that a symbolic convention such as the rule of signs—as it used to be called—leads to the generalized notion of quantity as positive or negative, in other words, to the notion of algebraic quantity.

That the import or significance of the algebraic law of association is clearly intelligible only if we start with the notion of quantity as positive or negative, and of the summation or aggregation of these quantities, is evident from the very nature of the explanation given of the way in which the convention which it embodies is reached; for the notion is at the outset introduced in one of the cases illustrative of it or subsumable under it: the case of credits and debits, as we shall presently see. In the meanwhile we may as well once more insist upon what is so familiar and obvious that it seems scarcely to challenge attention, viz. that these laws of association, distribution, &c., or conventional rules of manipulating the symbols of algebraic quantity, are themselves symbolic of certain processes of thought about algebraic quantity; and that it is because these invented manipulations have been found suitable for the purpose of symbolizing these mental processes that they have become conventions, are accepted as algebraical rules of expression.

This law of association is the fundamental convention of algebraic expression. What this convention is and how it is arrived at are explained in the second and third chapters of the *Introduction*, the second chapter dealing with the law of association only so far as addition and subtraction are concerned. I will give from this chapter two or three short extracts which touch upon points deserving of some attention.

'We shall suppose that $+a$ means a pounds to be paid by some debtor to a merchant A, and that $-b$ means b pounds to be paid by A to some creditor of his. It will facilitate matters if we suppose that A collects his debts and pays his creditors through an agent B, who may be supposed to have a certain amount of spare cash of his own. . . .'

This is with the object of calling attention to the fact that the order in 'the chain of additions and subtractions $+a+b-c+d-e-f$' is indifferent to the result; in short, expression is given to what algebraists somewhat magniloquently call the Law of Commutation. The symbols of addition and subtraction are here used as qualifying symbols, that is, to qualify the symbols of quantity—in this illustrative case sums of money; the symbolic expressions of qualified quantity are then formed into a chain or row, which is called an algebraic sum. We do not put these symbolic expressions into a chain or row by accident; the connecting them in that manner may be regarded as symbolic of a wider conception of aggregation; that is to say, the ideas of addition and subtraction are subsumed under a more general concept of aggregation, and this is plainly indicated in the choice of the term 'algebraic summation', or 'algebraic sum'.

Passing on to the derivation of the Law of Association, we find this explained as follows:

'Since two separate debts of £1 each, both supposed good, are from the merchant's point of view the same thing as a debt of £2, we may *associate* $+1+1$ into $+(+1+1)$, the bracket indicating that the two separate debts are regarded as one, and the

+ before the bracket meaning "payable to *A*" as before. We have therefore

$$+1+1 = +(+1+1) = +2;$$

and in like manner

$$+1+1+1 = +(+1+1+1) = +3;$$

and so on.'

This is, so far, only 'the simplest case of the Law of Association for addition'. What does the reader think of it, and of the reasons for adopting it? Remark that in 'the chain of additions and subtractions $+a+b-c$ . . .', which is called an algebraic sum, we have already associated these symbols, or there is no meaning in speaking of them as constituting a chain, or in calling them a sum. Why then are we not to regard the expression $+1+1$ as directly symbolic of the association of $+1$ and $+1$? If $+a+b$ signifies the sum of the credits £a and £b, and if $+1$ means a credit £1, why should not the expression $+1+1$ symbolize the sum of the credits £1 and £1; and hence why should we go out of our way to invent such an expression as $+(+1+1)$? The fact is that an expression such as this results from the application of the law of association in particular cases; does not serve to explain the genesis of that law, but rather to produce in the mind of the learner the impression that Algebra is the 'mere farrago of rules and artifices' into which, as Professor Chrystal remarks in his preface, 'the English textbooks of Algebra in vogue during the latter part of this century'—i.e. the nineteenth century—'have tended to degenerate.' A little further on we get the complete derivation:

'The process of association may be carried further. Let us suppose that *A*'s agent *B* in a day's round collects £a, pays out £b, and also pays out £c. Associating the whole of the day's business together, the result from *A*'s point of view is $+(a-b-c)$. If we look at it from the point of view of *B*'s cash, he owes to *A* £a, and *A* owes him £b and £c—that is to say, from *B*'s point of view *A* owes him $-a+b+c$, therefore, if we look at the whole result of the day's transactions again from *A*'s point of view, the result is $-(-a+b+c)$, the $-$ before the bracket meaning "due by *A*", as before. Combining the two results just arrived at with the original way of looking at each debit separately, we we have the following equalities:

$$+a-b-c = +(+a-b-c);$$
$$+a-b-c = -(-a+b+c).$$

'The last two equations exhibit fully the Law of Association for Addition and Subtraction. . . .'

There is in this development of algebraic expression something which reminds one of homonymy in ordinary language, and of the dependence of meaning of an homonymous symbol upon the context in which it stands. I think the recognition of this partial analogy is of some importance, and perhaps I shall make my meaning more clear if I suppose another way in which this law of association might suggest itself. Take, first, a very familiar concrete case. I wish to know how my current account at the bank stands, or rather how I stand financially in relation to my banker. I add up my debits on the one hand, my credits on the other, and take the difference between the two sums. Now, given the notion of quantities standing to each other in the relation of credit to debit, or merely this less general notion itself; also the artifice, suggested by this relation, of using the signs of addition and subtraction as qualifying symbols for such quantities—suppose then that I wish to express such a process as the above in the most general manner. It is natural, almost inevitable, that I shall be guided by the concrete familiar case and that I shall write

$$-(c+e+f)+(a+b+d), \quad . \quad . \quad . \quad . \quad . \quad . \quad . \quad . \quad . \quad . \quad . \qquad (A)$$

the two groups of debits and credits being indicated respectively by the *minus* and *plus* signs. But in such an expression lies the germ of the future rule, and a chance consideration, an accidental circumstance, may give it birth. For instance, let the supposition suggest itself of an error in the record of credits and debits: say that b has been wrongly entered as a credit in place of e. It will be at once obvious to the arithmetician that, so far as the result is concerned, the error can be corrected in two ways, either by exchanging b and e so that each is added to its proper group, or by subtracting each from the group into which it has been placed; so that, correcting the error in these two ways, we get

$$-(c+b+f)+(a+e+d) = -(c-e+f)+(a-b+d) \quad . \quad . \qquad (B)$$

This at once suggests the rule, viz. that the chain, row, or algebraic sum $+a-b-c+d+e-f$ . . . may be transformed or broken up into any two or more groups, provided that if we mark any group *plus* it shall contain only quantities marked with their proper signs, while if the group is marked *minus* it shall contain quantities only whose signs are reversed. If this rule of construction, or of grouping, is admitted, then the rule of interpretation, or of ungrouping, is at the same time determined, and we get, in fact, the rule of signs:

$$+(+a-b+c\ldots)=-(-a+b-c\ldots)=+a-b-c \quad . \ (C)$$

Now what I meant by the remark about homonymy and the way in which, in algebra as in ordinary language, the interpretation of an homonymous symbol necessarily becomes dependent on the content in which it stands, is this:

In the expression (A), before the rule of interpretation has developed, there is no ambiguity about the signs, it being intended and clearly understood that every quantity in the group marked $-$ is negative, and every quantity in the group marked $+$ is positive. Passing over the intermediate expression (B), in which the rule is in the making, to the final expression (C), in which the rule is embodied, we find that the signs $+$ and $-$, both inside and outside the brackets, have lost the distinctive meanings which they had in (A). The sign outside the bracket no longer denotes the character (positive or negative) of the quantities within it, and at the same time the signs ($+$ and $-$) within the brackets have lost their purely arithmetical meaning, for, e.g. all the quantities within a bracket may now be subtrahends, and this has no arithmetical meaning. Neither the sign prefixed to a quantity contained in a group, nor the sign prefixed to the group, determines the character of the quantity; it is determined by these two signs taken in conjunction or in their context; and we thus get three different ways of symbolizing both a positive and a negative quantity, that is, we have

$$+a=+(+a)=-(-a)\ ; \quad -a=+(-a)=-(+a)\ ;$$

and we may say that the first lot are all equivalent symbols of a quantity whose measure is a and whose sign is positive, while the second lot are all equivalent symbols for a quantity whose measure is a and whose sign is negative.

Returning to the connexion between the Law of Association and Algebraic Quantity as envisaged in this textbook, we see more clearly than ever that an inversion of the natural and logical order of thought is involved in the statement: 'The Law of Association for an algebraic sum, in particular the four special cases $+(+a)=+a$, $+(-a)=-a$, $-(+a)=-a$, $-(-a)=+a$, leads us to another important idea, viz. the notion of Algebraic Quantity as distinguished from what may be called mere Arithmetical Quantity.' The result of this inversion becomes manifest in the explanation of Algebraic Quantity which immediately follows:

'In the first instance, the operands a, b, . . . were mere numbers (e.g. numbers of pounds in our debit and credit illustration); but in the expressions $+(+a)$ and $+(-a)$ the operand as regards the first $+$ is not a, but $+a$ in the one case and $-a$ in the other. Such an operand, consisting of an arithmetical quantity with either $+$ or $-$ attached, we call an Algebraic Quantity, positive or negative according as the sign is $+$ or $-$.'

If we were to take this explanation literally, we should have to understand that 'algebraic quantity' is merely a technical name for symbolic combinations such as $+a$, $-a$, the meanings of these symbolic combinations being left unexplained. But I have no doubt that it is the meaning or interpretation of these combinations which the explanation is intended to give; that is to say, that $+a$ and $-a$ are symbols of positive and negative, or algebraic, quantity. According to the explanation, then, $+a$ and $-a$ are symbols of algebraic quantities, while $+(+a)$ and $+(-a)$ are symbols of algebraic quantities with signs of operation prefixed to them. If that is so, then $+a$ and $+(+a)$ are not identical in meaning, nor has $-a$ the same meaning as $+(-a)$. But then, what interpretation are we to give to the symbol $=$? Does not this symbol, in the equations $+(+a) = +a$, and $+(-a) = -a$, mean that the equated expressions are equivalent symbolisms? But if they are equivalent symbolisms how can they at the same time differ in meaning?

I note, next, the following passages:

'Strictly speaking, positive and negative quantities are not comparable as to magnitude, seeing that they are heterogeneous. Thus, for example, we cannot in the ordinary sense of the words say that "£5 due to *A*" is either greater or less than "£3 owed by *A*".

'It is usual, however, to establish a conventional test of inequality between positive and negative quantities by laying down that the algebraic quantity a is greater or less than the algebraic quantity b according as the reduced value of $a-b$ is positive or negative' (p. 19).

The first of these two paragraphs will hardly appear convincing to any one who reflects that we could and would say, in the given case, and in the ordinary sense of the words, that the sum due to A is greater than the sum owed by him; and that if we were asked, By how much greater? we would not hesitate to answer, Greater by £2. It is also quite a common thing to say that a man's assets exceed, or fall short of, his liabilities. If we consider that quantity, whether positive or negative (i.e. thought of in relation to measurable things or actions—'objects of measurement'—which, in given circumstances, evoke the notion of measure in opposed directions)—is still quantity; and that, from the point of view of abstract measure, quantity and magnitude cease to be distinguishable, we shall find it very difficult to agree with the above statement. For my part, I see in it an untenable position, to escape from which it is necessary to invent a fiction, a 'conventional test of inequality', the convention apparently consisting in the use of the ambiguous expression 'reduced value', coined, so it seems, in order to avoid the palpable inconsistency which would be involved in the use of the word subtraction in relation to two quantities said to be not comparable as to magnitude.

The algebraic symbolization of this conventional test of inequality between a positive and a negative quantity, and between two negative quantities, affords an instructive example of the inanities in symbolism, the mere playing with symbols, into which we lapse when we fail clearly to realize that the development of a symbolic system follows upon and expresses the development of conception, and that to invert this process, to suppose that development of conception waits upon or results from that of symbolism, merely leads (as we shall see presently) to the mystical supposition that the process of conceptual development is, or ought to be, something other than we actually find it. It will be sufficient to consider the algebraic symbolization of the test in the case of its application to the comparison of a positive with a negative quantity:

'If a be any positive quantity, say $+\alpha$, b any negative quantity, say $-\beta$ (here α and β are absolute quantities), then $a-b = +\alpha-(-\beta) = +\alpha+\beta$, and obviously has a positive value. Hence any positive algebraic quantity, however small absolutely, is greater than any negative algebraic quantity.'

In a similar way it is shown that—

'one negative quantity is greater or less than a second according as the first is absolutely less or greater than the second' (p. 19).

In case it should not be quite clear what I mean by calling this an inanity in symbolism, or a play upon symbols, I will explain. But before I do so I should like to go back to the underlying idea that positive and negative quantities are not strictly comparable as to magnitude, so that we are led to invent a conventional test of inequality between them. The point is, how does this idea of heterogeneity between positive and negative quantities arise? Does the assertion really correspond with the mental fact, and is the conventional test anything but an endeavour to establish this correspondence by a play upon words? We cannot but remark that this assertion about positive and negative quantity in the abstract is in obvious conflict with the way in which we think and speak of positive and negative quantities in particular and in the concrete. We certainly do think about, and say of, opposed forces, velocities, accelerations, momenta, &c., that they (each of their kind) are equal to, greater or less than, one another; we say, of an excess above a mean, that it is equal to, greater or less than, a defect from the mean; we say that a length, measured in one direction from a point of origin, is equal to, greater or less than, a length measured in the opposite direction from the point of origin; some of us think and say that a credit is equal to, greater or less than, a debit—and I think it not improbable that the mathematician thinks and says so too when he is not hampered by a conventional test of inequality. What, then, can the motive be which prompts mathematicians to say, of positive and negative quantities in general, or in the abstract, that which they do not appear to think, and certainly do not usually say, of positive and negative quantities in particular, or in concrete cases? It is, perhaps, this: that if we were to admit equality as a possible relation between positive and negative quantities we should have to admit such algebraic expressions as $+a = -a$, which would bring the system into utter confusion and entail a complete reconstruction of it from the very beginning. I can think of no other motive sufficiently imperative to prompt a statement so difficult to reconcile with the actuality of our conceptions and the phraseology in which we habitually express them. But if this statement is to be thus accounted for, it merely affords a striking example of the tyranny which symbols exercise upon us when we cease to regard them from a purely rational standpoint, and fall into the mystical attitude. Such a conflict as this, were it real, between the symbolism of algebra and the nature of the conceptions which give it birth, would be the condemnation of the former. But there is in reality no such conflict, and no necessity to seek refuge from it in a conventional test of inequality. A mathematical symbol, like any other symbol, has that meaning, or those meanings, which we please to assign to it; its sense is the sense in which we actually use it. Well, what is the use to which mathematicians actually put the symbol =, which they define as 'is equal to'? To denote, not mere equality, but identity. Hence, as neither they nor any one else thinks of a positive quantity as identical with a negative quantity, they cannot admit $+a = -a$ as a valid expression of their thought. But while they actually use the symbol = as a symbol of identity, they customarily *term* it a symbol of equality; hence the apparent, but quite unreal necessity of concluding that there cannot be equality between positive and negative quantities, that these quantities are not properly comparable as to magnitude, and that the fact that we continually do thus compare them is to be ignored, and something which is called a conventional test of inequality substituted for it.

Although I believe it cannot fail to become clear to any one who reflects upon the principles of algebraic symbolism that the sign =, though commonly referred to as a sign of equality, is in reality always used in algebra as a sign of identity (which of course includes equality), yet I may as well reinforce this position by quoting the mathematical expert himself. If the reader will refer to pp. 46 and 47 of the textbook, he will see that equations are there classified as Identical and Conditional; and he will also see that a conditional equation is a conditional identity, although this is not stated in so many words. The examples given make this quite clear:

'For example, $(x+1)(x-1) = x^2-1$, and $3\times 2+2\times 2 = 5\times 2$ are identities; but $2x-3=x+1$, and $x+y=x^2+y^2$ are conditional equations'—that is, conditional identities. Then, commenting on the use which some writers make of the signs = and ≡ in order to distinguish between these two cases, the author states that he adheres in general to the old usage for a variety of reasons: 'Chief among them is the view which will be found to pervade this book, that all algebraic equality (which is not approximate) is, at bottom, identity. The same is true of arithmetic equality. Algebra is, in short, "the Calculus of Identity"' (footnote, p. 47).

There is, then, nothing in algebraic symbolism to induce us to shut our eyes to the evidence which our conceptions carry in themselves. Any one who finds that the relation between positive and negative quantity is in no way different in the abstract from what it is in the concrete, in the general from what it is in the particular, may admit this without the least fear of thereby damaging the validity of algebraic symbolism, and may proceed in the even tenor of his way with complete indifference to conventions which embody no real process of thought. And this brings us back to the consideration of the conventional test in question as expressed in algebraic symbolism. I averred this process to be equivalent to what, in ordinary language, would be a mere play upon words; and I think this describes it with very tolerable accuracy. Let me repeat here the paragraph already quoted: 'If a be any positive quantity, say $+\alpha$, b any negative quantity, say $-\beta$ (here α and β are absolute quantities), then $a-b=+\alpha-(-\beta)=+\alpha+\beta$, and obviously has a positive reduced value. Hence any positive algebraic quantity, however small absolutely, is greater than any negative algebraic quantity.'

Now this paragraph follows almost immediately upon the explanation given of the difference between arithmetical and algebraic quantity and their respective symbols. In effect, then, it amounts to this: If, ignoring the symbolic distinction just given between arithmetical and algebraic quantity, we make a and b symbols respectively of a positive and of a negative quantity; and, at the same time admitting this symbolic distinction, we symbolize these quantities by $+\alpha$ and $-\beta$, then we may write $a=+\alpha$, $b=-\beta$, and it will follow that $a-b=+\alpha-(-\beta)=+\alpha+\beta$.

Certainly; but while we are about it we may as well strip the process of its transparent disguise, and, as before, at once ignoring and recognizing the said symbolic distinction, write $a=+a$, $b=-b$, whence it will follow that $a-b=+a-(-b)=+a+b$, an equation which is perfectly legitimate under the condition that a and b in the left-hand member respectively symbolize a positive and a negative quantity, while in the other members they symbolize arithmetical quantities. Or, again, we may put it that in the left-hand member the qualificative signs of the quantities are implied or understood. Make explicit what is implied and write $+a-(-b)=+a-(-b)$. All this is mere play with symbols; but, as thus put, it is play open and undisguised, and we are under no illusion that a process of conception is thus being developed.

Now in whatever way the alleged notion of heterogeneity between positive and negative quantity may have arisen, and whatever may be the connexion between this notion and the conventional test of inequality, and between the conventional test and the conclusions that any positive quantity, however small, is greater than any negative quantity, and that one negative quantity is greater or less than a second according as the first is absolutely less or greater than the second—if these conclusions are accepted (that is, accepted verbally and in the belief that they have a definite conceptual content), there will follow the further convention in symbolism:

'If, therefore, we use ∞ to mean a quantity greater than any assignable quantity, then we may symbolize the whole series of algebraic quantity by

$$-\infty \ldots -2 \ldots -1 \ldots \pm 0 \ldots +1 \ldots +2 \ldots +\infty$$

the order of ascending magnitude being from left to right.'

That is, we are supposed to have arrived at a general conception of algebraic quantity, conventionally expressed in the above manner, the characteristic feature of which lies in the ascending order of magnitude from $-\infty$ to $+\infty$; positive and negative quantity starting from no quantity and respectively growing more and more, and less and less. Any one who can recall his schooldays, in particular his initiation into the mysteries of algebra will, I doubt not, also recall the bewilderment produced in his mind by the authoritative divulgation of quantities less than no quantity and infinitely less than no quantity: a bewilderment which gradually yielded to the lethal effect of a sufficiently oft-repeated formula, accepted as significant with the trustfulness natural to youth and ignorance at the bidding of the pastor and master.

Surely that cannot be essential in algebraic convention which obliges us to disregard clearness of conception and cast a slur on common sense. 'Quantity' is the name of a relation the conceiving of which is in no way modified when we combine with it that of opposition in measure between the 'things' quantitatively related; and the case is in no wise altered when, the nature of the 'things' becoming indifferent to the purpose of our thought, we make abstraction of them altogether and confine the subject of thought to this abstract combination of quantitative relation and opposition in measure. This amounts to saying that the term 'algebraic quantity', or its equivalent in algebraic symbolism, expresses this abstract combination, and not any modification or extension of the idea of quantity. From this point of view, which at least does no outrage to common sense, such a phrase as 'the series of algebraic quantity extends in ascending order of magnitude from $-\infty$ to $+\infty$ is a phrase and nothing more: it corresponds to no real process of conception.

Let us return now to the rule of signs and its illustration by means of credit and debit. As before, let $+a-b+c-d+e-f+g-h$ symbolize a summation of credits and debits in terms of any monetary unit A. In accord with the rule of signs we may break up this sum into the group-form

$$+(+a-b)-(-c+d)+(+e-f)-(-g+h) \quad . \quad . \quad . \quad (1)$$

Now suppose the credits and debits are to be expressed in terms of a monetary unit the nth part of A. Then we might write (1) thus: $+(+n \times a - n \times b) -$ &c. But by convention we may more briefly write it thus:

$$+n(+a-b)-n(-c+d)+n(+e-f)-n(-g+h) \quad . \quad (2)$$

Now if in (1) we suppose $a > b$, $c < d$, $e < f$, $g > h$, and symbolize these differences in their order by $+p$, $+q$, $-r$, $-s$, we shall get (1) in the form

$$+(+p)-(+q)+(-r)-(-s) \quad . \quad . \quad . \quad . \quad . \quad . \quad . \quad . \quad . \quad (3)$$

and we shall get (2) in the form

$$+n(+p)-n(+q)+n(-r)-n(-s) \quad . \quad . \quad . \quad . \quad . \quad . \quad (4)$$

But since (3) is by the rule of signs equivalent to $+p-q-r+s$, and since the balance, whether credit or debit, must in (4) be measured by a number n times that by which it is measured in (3), we must admit, as a legitimate result of the rule of signs and the above convention, the equation

$$+n(+p)-n(+q)+n(-r)-n(-s) = +np-nq-nr+ns.$$

Thus, generalizing, $+n(-p)$, $-n(+q)$, $+n(-r)$, $-n(-s)$ are, in virtue of the said rule and convention, symbolic expressions of quantities whose measures respectively are $n \times p$, $n \times q$, $n \times r$, $n \times s$, and whose signs respectively are $+$, $-$, $-$, $+$. It will be remarked that we arrive at this result without using any such phrases as 'the multiplication of one algebraic quantity by another algebraic quantity', or 'the multiplication of an algebraic sum by an algebraic quantity', or 'positive and negative multiplication'—whatever such phrases may mean. We have been concerned simply with two conjoined but quite distinct operations, viz. multiplication and sign-determination in accord with the rule of signs.

The operation of the rule of signs in conjunction with that of multiplication is explained in chapter iv of our textbook, under the title 'The Law of Distribution'. I naturally presume that explanation to be in accord with the views of mathematicians. It betrays, in my opinion, clear evidence of a fresh lapse into the region of mysticism. I need not go beyond the first page of this explanation in order to make my meaning plain:

'The primitive meaning of multiplication is repeated addition. Thus 8×3 is a contraction for $8+8+8$.

'In our discussion of the laws of commutation and association for multiplication and division we considered only the case where the operands are absolute, i.e. merely arithmetical quantities. The further points that arise when the operands are algebraical quantities—that is to say, absolute quantities with the signs $+$ or $-$ attached—are most conveniently considered in connexion with the Law of Distribution, which is the last of the three fundamental laws of algebraical operation.'

Let me here once again emphasize the absolute necessity, in expounding the principles of any system of symbolizing thought, of never allowing the distinction between the process of thought and the process of symbolization to lapse from the mind, however convenient for brevity's sake it may be to refer to the symbols as if they were one and the same with that which they are intended to symbolize. Once the principles are understood and agreed to, the necessity is no longer imperative but incidental.

'Reverting to 8×3, let us write the product more fully as $(+8) \times (+3)$, and notice that we may also write $8+8+8$ more fully in the form $+8+8+8$, or if we choose $+8 \times 1$ $+8 \times 1$

$+8\times 1$. Remembering that $+3$ is a contraction for $+1+1+1$, we may therefore write the equation $8\times 3 = +8+8+8$ in the forms—

$$(+8)\times(+3) = +(+8)+(+8)+(+8)$$
$$= +8+8+8 = +24 \quad . \quad . \quad . \quad (1);$$
or
$$(+8)\times(+1+1+1) = +8\times 1+8\times 1+8\times 1 \quad . \quad . \quad (2).'$$

It follows of course from this that $8\times 3 = +24$, and that $24 = +24$. It follows, in short, that if we abolish the distinction in symbolism between arithmetical and algebraic quantity we abolish that distinction. This is not an instructive explanation, even though it be presented as an example of 'multiplication by a positive multiplier'. There is obviously here an apparently gross inconsistency with what has gone before. After carefully elaborating the distinction between arithmetical and algebraic quantity and the correlative distinction between their symbols, this distinction is summarily abolished by writing 8×3 'more fully' as $(+8)\times(+3)$! We may ask: Since you abolish the symbolic distinction between arithmetical and algebraic quantity why not as well write 8×3 'more fully' as $(-8)(+3)$, or $(+8)(-3)$, or $(-8)(-3)$? I suppose Professor Chrystal would answer, with Cayley, that 'the numbers or magnitudes which we represent by symbols are in the first instance ordinary (that is, positive) numbers or magnitudes';[1] and that the laws of algebra subsequently lead us to introduce 'essentially negative' quantity.[2] Thus 8×3 would be merely an abbreviation of $(+8)(+3)$. But this is obviously inconsistent with the distinction previously drawn between arithmetical quantity on the one hand and algebraic, i.e. positive and negative, quantity on the other; it could only be maintained by making ordinary or arithmetical quantity = positive quantity, and algebraic quantity = negative quantity.

I am unable to see, in this opinion of Cayley's, apparently shared by Professor Chrystal, anything but a sheer illusion as to the nature of the conception of quantity, an illusion explicable only, it seems to me, as the result of an unperceived reaction of the process of algebraic symbolism upon the process of thought. Quantity, in the first instance, and in the last, is no more positive than it is negative. The terms 'positive'

[1] See p. 82 (On Cayley's Presidential Address).

[2] *Introduction to Algebra*, p. 184.

and 'negative' are correlative; and consequently it is not clear what significant purpose can be served by calling quantity positive (or negative) save in express contradistinction to calling it negative (or positive).

It would be tedious, and it is unnecessary, to pursue this criticism of 'algebraic multiplication' throughout its development as given in our textbook; but the point of the criticism may perhaps be made still more clear if we consider one more, the next, step in this development, viz. simple 'negative' multiplication. This is explained as follows:

'We thus look upon multiplication by a positive multiplier as a contraction for repeated addition. In like manner, it is natural to regard multiplication by a negative multiplier as a contraction for repeated subtraction. Taking this view, we have—

$$(+8)\times(-3) = -(+8)-(+8)-(+8)$$
$$= -8-8-8 = -24.\text{'}$$

It may be that if any one conceives repeated addition as 'positive' multiplication he will also be able to conceive repeated subtraction as 'negative' multiplication. But why should we not as well reason in the following manner: Multiplication and division are inverse operations. To multiply 8 by 3 is to add 8 and 8 to 8, which gives 24 as the product of 8 by 3. The inverse of this operation is to divide by 3 this product 24, or to subtract 8 and 8 from 24, which brings us back to 8, the quotient of 24 by 3. Now $(+8)\ (+3)$ being 8×3 written 'more fully', so that '$(+8)\ (+3) = +24$' means the same as '$8\times 3 = 24$', and division or repeated subtraction being the inverse of multiplication or repeated addition—then if 'it is natural to regard multiplication by a negative multiplier as a contraction for repeated subtraction', it will follow that the inverse of the process symbolized by $(+8)\ (+3) = +24$, that is, the division of 24 by 3, must be symbolized by $(+24)\ (-3) = +8$. Hence also, if this operation, commonly written $\frac{24}{3} = 8$, be likewise written more fully $\frac{+24}{+3} = +8$, we get $\frac{+24}{+3} = (+24)\ (-3)$ and, in general, $\frac{+a}{+b} = (+a)\ (-b)$.

This reasoning is neither more nor less cogent than that in the text criticized; and neither the one nor the other has any real value because they both proceed from the same vicious reaction of symbolic forms upon the conceptual process. The expressions 'positive' multiplication and 'negative' multiplication have no literal meaning, that is, they do not correspond to any modification of the conception of multiplication or repeated addition; and to found an argument on a verbal implication which does not correspond with the mental facts is to fall into mysticism.

I can understand the multiplication, or repeated addition, of an arithmetical quantity; the multiplication, or repeated addition, of a positive quantity, or of a negative quantity; but the 'positive' and 'negative' multiplication of any of these quantities has for me no meaning at all. There is, of course, no reason why we should not use the expression 'algebraic multiplication' as a technical term or brief way of indicating a dual operation, viz. multiplication and sign-determination; and I am quite prepared to be told that I am forcing an open door—that this is, in fact, the sense in which this expression is understood by algebraists. But if that is the case why do they not plainly state it instead of concealing it within a maze of confused and ambiguous phraseology; and why do they give explanations of this process which obscure, instead of making plain, its real nature?

I am far from affirming—it would be impertinence on my part to affirm—that there are not mathematicians who take a rational view of the process of algebraic symbolism in general; but I think these ambiguities of phraseology, and these explanations which do not explain, are in themselves evidence of a lapse from the rational to the mystical, of a reaction of symbolism upon the process of conception which, not being realized, produces illusion of judgement as to the real nature of that process. In ultimate analysis this appears to be due to the simple fact that we find the expression of thought ready made for us, and, instead of having to originate symbolism (an effort which involves vivid awareness of the distinction between expression and that which is expressed), merely learn it, and, in large measure, learn to think in learning to speak. Thus the two processes often seem to be one and the same, and some persistency of effort is required adequately to realize the duality and the nature of the interdependence. It is in the persistence of this effort that we come to apprehend the conditions of the problem of expression as it presented itself to the originators of any system of symbolism and to those who successively perfected and developed it; and it is also in the persistence of this discrimination that we avoid falling into a mystical attitude of mind.

But this is not the way in which at present the beginner is taught algebra, and the teacher himself seems to have learnt in no other way than the way in which he teaches. The ineptitudes which, even now, after all the efforts at reformation made by Professor Chrystal and many others, characterize the orthodox exposition of elementary algebra, appear to be in the main due to an insufficiently clear and pervading sense of the difference between thinking and symbolizing, and of the mode of action of a system of symbolism in the development of a process of thought. In every one of the illusory judgements to which we have called attention we can trace the retro-active and distorting pressure of the symbolism used. To return to the case we have just had in review, that of 'algebraic multiplication'. Expressions of the type $(+a)\times(+b)$, $(+a)\times(-b)$, &c., are introduced to the learner not as expressions which result inevitably from the application of conventions previously explained, but as symbolic forms giving expression to an extension of the ordinary or primitive concept of multiplication. He has learnt that $+a$, $-b$, &c., are symbols of algebraic quantity; he finds them here conjoined with the symbol of multiplication; and he is told to read the combination as symbolizing the conception of an operation which, in ordinary language, is described as the multiplication of one algebraic quantity by another, or as positive and negative multiplication of a positive quantity, and of a negative quantity. His attention is not recalled to the fact that in the very adoption of the antecedent conventions for grouping without ambiguity symbols of algebraic quantity, the qualifying signs $+$ and $-$ lose the definite meaning which they have in isolated expressions such as $+a$, $-b$, or in a simple chain or sum of such symbols, so that, in such group-formations, the qualification of any quantity whose measure is, say, n, is only unambiguously symbolized by a context or conjunction of the signs $+$ and $-$; in other words, that in such group-formations we cannot regard a combination such as $+a$ or $-b$ as symbolic respectively of a *positive* quantity whose measure is a, or of a *negative* quantity whose measure is b, so to regard them being plainly inconsistent with the conventions adopted. Thus when we get such an expression as $(+a)(-b)$ and, yielding to the suggestion which the very form of it conveys to the mind, we look upon the $+a$ and the $-b$ contained in it as respectively symbolizing a positive and a negative quantity, we misinterpret the artifices of expression which we have ourselves invented or adopted, and this misinterpretation in its turn reacts upon and distorts the judgement we make of the process of thought which suggested them. Instead of interpreting $(+a)(-b)$ as symbolic of a quantity whose measure is the product $a \times b$ (number being the measure of purely abstract quantity), and whose qualification is, according to the rule of construction and interpretation adopted, negative, we say that it symbolizes the multiplication of a positive quantity by a negative quantity, or vice versa, and then proceed to invent explanations of this new kind of multiplication.

The fact is, as we have seen, that such equivalences as $+ab$, $(+a)(+b)$, $(-a)(-b)$; $-ab$, $(+a)(-b)$, $(-a)(+b)$ inevitably arise in the application of the admitted conventions, and are nothing else than the different ways in which, consistently with these rules, a multiple quantity, positive or negative, may find itself expressed. We should not allow ourselves to be hypnotized by the mere form of these expressions into the belief that we conceive some sort of process literally definable as the positive or negative multiplication of a quantity either positive or negative.

But if, instead of a rational, we take a mystical view of this dual process which is called algebraic multiplication; that is, if the judgement we pass (either explicitly or implicitly) upon the conceptual process which prompts the symbolization, involves a fictitious element of which we are unaware, then we shall evidently come to the consideration of algebraic 'roots' and 'powers' under bias of this mystical view, and the fictitious element will reappear in the judgement of the thought-process symbolized by these forms. If this is the case the question at once suggests itself whether this illusion in judgement is not in some way connected with the 'notion' (belief?) derived, according to Cayley, from Harriott's symbolization of an equation in the form $f(x) = 0$, that an equation of the order n ought to have n roots; which in turn leads to 'the notion which is really the fundamental one . . . underlying and pervading the whole of modern analysis . . . that of imaginary magnitude. . . .'

It is this question, and the further analysis of elementary algebraic forms to which it gives rise, which constitute the subject-matter of the next chapter.

CHAPTER VIII

THE CONCEPTIONS AND SYMBOLISM OF ELEMENTARY ALGEBRA (*continued*)

Analysis of the relations implied by the use of the correlative terms Power and Root in Arithmetic and in Algebra.—The actual use of the power-index is inconsistent with the implied definition of Power and Root in Algebra.—But this inconsistency is convenient because it confers brevity and symmetry upon the symbolic system; and its real sanction lies in its arithmetical interpretability.—The convention, once admitted, leads by strict analogy to a similar use of the root-index, and suggests the pseudo-concept of Imaginary Quantity.—The textbook explanation of Imaginary Quantity.—The sophisms which this explanation involves.—Recapitulation of the two ways of interpreting the development of symbolism in Elementary Algebra.—Argand's geometrical representation of Imaginary Quantities.

'In addition to the four species (of arithmetical operation) it is usual, even in arithmetic, to introduce another pair of mutually inverse operations, viz. Involution (Raising to a Power), and Evolution (Radication or Root Extraction). In the first instance, at least, these new operations are not independent of those already enumerated. Involution is, in fact, repeated multiplication: thus $3, 3\times3, 3\times3\times3, 3\times3\times3\times3, \ldots$ are represented by $3^1, 3^2, 3^3, 3^4, \ldots$ and are described as three to the first power, three to the second power or three square, three to the third power or three cube, three to the fourth power, . . . and in general $a\times a\times a \ldots$ (n factors), n being of course an arithmetical integer, is contracted into a^n. . . . The quantity whose nth power is a is called the nth root of a, and is denoted by $\sqrt[n]{a}, \ldots$' [1]

So far as arithmetical quantities are concerned we have here a clear explanation of the numerical relations which are intended to be conceived under the correlative terms 'root' and 'power'. But I am unable to find, anywhere in the textbook under consideration, an explicit definition of the relations intended to be conceived when the terms 'root' and 'power' refer to algebraic quantity. I suppose, however, that the definition would run somewhat as follows: An algebraic quantity is termed a power in relation to another algebraic quantity, which is termed

[1] *Introduction to Algebra*, p. 5.

a root, when the former arises from the involution of the latter; or, conversely, an algebraic quantity is termed a root in relation to another algebraic quantity, which is termed a power, when the former arises from the evolution of the latter.

This definition, however, is of no use to us unless we know the exact sense in which the words 'involution' and 'evolution' are used in relation to algebraic quantity. But if, in order to find this out, we turn to the Index of Technical Words and Phrases given in the beginning of the book, we find ourselves referred to p. 5, that is, to the explanation already quoted, in which the meanings of the terms 'involution' and 'evolution' are explained only in relation to arithmetical quantity.

In these circumstances we must suppose that what algebraists precisely conceive under the correlatives 'root' and 'power', when these terms refer to algebraic quantity, is intended to be revealed by the study of the contexts in which these terms appear, by the use to which they put indices or exponents of power and the radical sign, and by the manner in which they combine these new conventions with the others, especially with the rule of signs. Let us take the simplest examples possible, and, using the current phraseology of algebraic exposition, consider how far they reveal a clear process of conceiving and of symbolizing.

First, according to the convention of indices, so far as actually explained, we have

$$a \times a = a^2 \quad . \quad . \quad . \quad . \quad . \quad . \quad . \quad . \quad . \quad . \quad . \quad (1)$$

The same convention, if we extend it to 'algebraic quantities' gives

$$(+a)\,(+a) = (+a)^2 \quad . \quad . \quad . \quad . \quad . \quad . \quad . \quad . \quad (2)$$

$$(-a)\,(-a) = (-a)^2 \quad . \quad . \quad . \quad . \quad . \quad . \quad . \quad . \quad (3)$$

Next, by the rule of signs, we have

$$(+a)\,(+a) = +(a \times a) \quad . \quad . \quad . \quad . \quad . \quad . \quad . \quad (4)$$

$$(-a)\,(-a) = +(a \times a) \quad . \quad . \quad . \quad . \quad . \quad . \quad . \quad (5)$$

$$(+a)\,(-a) = -(a \times a) \quad . \quad . \quad . \quad . \quad . \quad . \quad . \quad (6)$$

Now it is the actual practice of algebraists to equate (4), (5), and (6) respectively to $+a^2$, $+a^2$, $-a^2$. I suppose the explanation of this to be that by equation (1) we may substitute a^2 for $(a \times a)$ in (4), (5), and (6). But it will be seen, by comparing (2) and (3) with (4), (5), and (6), that in effecting this substitution we introduce an ambiguity in the use of the index. That is to say, an index being an exponent of power, and the terms power and root being used as correlatives, we fail so far to form a definite and stable conception of the relation implied by these correlative terms as applied to algebraic quantity. For in (2) and (3), constructed analogically from (1), the index 2 clearly refers to the number of 'algebraic quantities' $(+a)$ or $(-a)$; but in (4), (5), and (6), when equated to $+a^2$, $+a^2$, $-a^2$, the index appears to refer to the number of arithmetical quantities a, and in (6) certainly cannot refer to two 'algebraic quantities', $+a$, or $-a$.

Now if, on the one hand, it is agreed that the numbers 1, 2, 3, . . . n, employed in this particular manner, shall indicate 1, 2, 3, . . . n identical 'algebraic quantities multiplied together', or, in other words, shall be indices of the successive orders of power of the identical quantity or root—an agreement which definitely fixes the relation intended to be conceived in the correlation of the terms 'root' and 'power', as predicated of algebraic quantity—then it is inconsistent to transform the expression $(+a)(-a)$ into $-a^2$, since the number 2, thus used, is by definition an index or exponent of power, that is, designates an algebraic quantity which stands to some other algebraic quantity in the relation of power to root, which is not here the case, since $+a$ and $-a$ are not identical. If, on the other hand, it be agreed that integers, thus used, are to indicate the number of arithmetical quantities multiplied together, then the relation conceived under the correlated terms 'power' and 'root' is an arithmetical relation, and it is meaningless to speak of an algebraic quantity as a power or as a root.

But algebraists do speak and write, throughout algebra, of roots and powers as algebraic quantities; thus, notwithstanding the inconsistency which attends their employment of indices, we cannot doubt that the relation which they conceive between two algebraic quantities which they respectively term root and power is that indicated in the first of these alternatives. But let me add here that I am far from asserting that a use of indices which is inconsistent with the nature of the relation conceived is necessarily a vicious use. It need not be so, provided we are clearly aware of the inconsistency, and deliberately accept it for the sake of the brevity and symmetry it confers upon the system. The danger to clear thinking, and the tendency to pseudo-conceiving, lie in not being aware of the inconsistency, or in shutting our eyes to it, for this is to remain defenceless against a prejudicial reaction of symbolic forms on the process of conception.

We may at this point pause to recall for a moment Mr. Whitehead's explanation of the enigma involved in the doctrine of imaginaries in algebra—especially the view that the laws of algebra, though suggested by arithmetic, are independent of it, and that once algebra is conceived as an independent science dealing with the relations of certain marks conditioned by the observance of certain conventional laws, the difficulty vanishes. In my criticism of this view I remarked that so to conceive algebra is just as great a difficulty as the difficulty it is supposed to remove, because no intelligible account can then be given of the invention, the *raison d'être*, of the relations, conditions, and laws or conventions affecting the symbols or marks. But it seems to be at least involved in this conception of algebra that its laws develop in self-consistency, however obscure the motive or cause of this development may be. The inconsistency in development which has just been pointed out seems, then, to be absolutely fatal to this view. The symbolism of algebra develops here in defiance of logical consistency; and the real sanction of this inconsistency in symbolism plainly is that it is *arithmetically* intelligible or interpretable.

The dilemma in which the orthodox exposition of root and power leaves us is quite clear. We cannot escape from the choice between two equally embarrassing conclusions. The correlative terms root and power either are or are not intended in algebra to indicate a relation which is different from that indicated in arithmetic by this correlation. While the orthodox exposition makes it clear that the correlation intended is not simply arithmetical, but algebraic, we cannot admit it without at the same time condemning as illogical the actual use of indices or exponents of power. On the other hand we can regard the actual use of exponents of power as logical, but we can then admit no difference between arithmetical and algebraic root and power, and it becomes meaningless to say that $+a^2$ has two roots and $-a^2$ none, and absurd to go on to invent two roots for $-a^2$. Here again we have no option but to abandon the orthodox, which is also the mystical, standpoint, and return to common sense. But we may as well first consider briefly the use to which algebraists put the radical sign $\surd$.

'The quantity whose nth power is a is called the nth root of a, and is denoted by $\sqrt[n]{a}$, a being called the Radicand, and n the Order of the Root; and the operation of deriving $\sqrt[n]{a}$ from a is called Evolution, Root Extraction, or Radication; special cases are the second root or square root, written $\sqrt{a}$; the third root or cube root, written $\sqrt[3]{a}$. If $a=b^n$, b being any ordinary arithmetical quantity, it is at once obvious that b satisfies the definition of $\sqrt[n]{b^n}$. It also follows from the definition that the nth power of $\sqrt[n]{a}$ is a. From these remarks the mutual inverseness of Evolution and Involution, regarded as arithmetical operations, follows at once.'[1]

We note, first, that b being any ordinary arithmetical quantity, b satisfies the definition of $\sqrt[n]{b^n}$. That is to say, $\sqrt[n]{b^n} = b$. Thus $\surd a^2 = a$; and, since $a \times a = a^2$, we also have $\surd a^2 \times \surd a^2 = a^2$.

In combination with the rule of signs we then get:

$$+(a \times a) = +a^2;\ \text{hence}\ +(\surd a^2 \times \surd a^2) = +a^2;$$
$$(+a)(+a) = +a^2;\ \text{hence}\ (+\surd a^2)(+\surd a^2) = +a^2;$$
$$(-a)(-a) = +a^2;\ \text{hence}\ (-\surd a^2)(-\surd a^2) = +a^2.$$

Now $+a^2$ is a 'square' or power of the second order, and the square roots of $+a^2$ are $+a$ and $-a$. Hence by the last two identities $+\surd a^2$ and $-\surd a^2$ also symbolize the square roots of $+a^2$. That is, using the radical symbol, $\sqrt{+a^2} = +\surd a^2$ or $-\surd a^2$.

Again, and for similar reasons, since we have the inconsistent expression $(+a)(-a) = -a^2$, so, to be consistent in inconsistency, we must also have $(+\surd a^2)(-\surd a^2) = -a^2$.

It is needless to repeat again here, and in detail, what was said in the last chapter on algebraic multiplication. I shall be understood if I simply reiterate the statement that in an expression of the type $(+a)(-b)$ . . . the combinations $+a$, $-b$, &c., are not symbolic of algebraic quantities. We may call them—and they already are so called—'factors' of a complex expression for algebraic quantity. In the particular case where the factors are identical, if the terms root and power are used, we shall consider these words as technical terms respectively for the repeated factor and for the complex expression, no matter in which of its equivalent forms the latter may actually be written.

[1] Ibid., p. 5.

It is significant, then, and also true, that $+aa$, or $+a^2$ is a power which has two roots, $+a$ and $-a$, for this statement means nothing more than that, in accordance with the rule of signs, $(+a)(+a)$ and $(-a)(-a)$ are equivalent expressions for $+aa$, or, according to the convention of indices, for $+a^2$.

But we remain with the actual inconsistent use of power exponents which gives us such expressions as $-a^2$, $-a^4$, $-a^6$, ... $-a^n$ (where n is an even integer), so that we have a defective correlation of power and root, or the absurdity of powers which have no roots, powers which are not powers. We could of course get rid of this absurdity by going back upon our steps and refusing to employ these inconsistent expressions; but as this would mean loss of symmetry, of brevity, and of convenience, and as symbolic systems are made for man, not man for symbolic systems, it is quite certain we shall do nothing of the kind. Nevertheless, in a science whose especial boast it is to be rigorously logical, the presence of an absurdity or inconsistency will with difficulty be tolerated by its votaries and adepts. If it is possible to invent any fresh artifice which shall afford a logical justification of such expressions as $-a^2$, $-a^4$, ... that is, if they can be turned into powers by being supplied with roots, the artifice is certain to be adopted, provided it is found not to conflict with already established artifices. As we are here taking a common-sense view of algebraic symbolism, we shall at the same time recollect that in the devising of such an artifice the roots to be invented need not, any more than any other roots, be symbolic of algebraic quantities.

Let me suppose a belated algebraist who has never even heard of imaginary quantity; who attaches clear and definite meanings to the terms root, power, and index or exponent of power with reference to algebraic quantity; who is quite aware of the inconsistency involved in the extended use of indices to cases which do not fall under his concept—the ordinary algebraic concept—of the relation of power and root, but admits this inconsistent use for the sake of the greater convenience, brevity, and symmetry which it confers on algebraic symbolism. Suppose now that such a man, ruminating over the advantages attending this extended use of the power-index, should ask himself whether some further accession of convenience or symmetry might not perhaps result from an analogous extended use of the root-index.

I can imagine him reasoning somewhat as follows: If for the sake of brevity and symmetry we admit such a combination as $-a^2$, that is, if we here employ the power-index 2 notwithstanding that $-a^2$ does not symbolize an algebraic power, is there any valid reason why we should not generalize the procedure by admitting such a combination as $\sqrt{-a^2}$; that is, use the root-index 2 just as we have used the power-index 2, $\sqrt{-a^2}$ not symbolizing a root any more than $-a^2$ symbolizes a power, and hence write $(\sqrt{-a^2})(\sqrt{-a^2}) = -a^2$?

Of course I should, personally, put the matter somewhat differently, since I regard the terms power and root in algebra as purely technical, not as indicating relation between one algebraic quantity and another; but the question as it thus stands might be asked, to-day, by any one who had an elementary knowledge of algebra, and had not let himself be persuaded that he conceives some new kind of quantity which is called 'imaginary quantity' by those who have invented it. And remark that the answer to this question is, from a philosophical point of view, of very little importance compared with the fact that the question itself is a perfectly rational, plain, and straightforward one. Without doing violence to common sense, without performing any *salto mortale* in the process of conception, we are brought to inquire whether we may not logically admit a fresh combination of symbols, the import of which, from the orthodox standpoint, is the subject of obscure, incoherent, and at times even self-contradictory explanation. So far as we are here concerned with the answer to the question, as put by the imaginary algebraist, all we need remark is that the proposed use of the root-index is in strict analogy with the previously admitted use of the power-index, and that the new artifice thus combines logically with those which precede it. The objection that $\sqrt{-a^2}$ does not symbolize an algebraic quantity might be awkward for the imaginary algebraist to meet, save after the orthodox manner, but it will leave those unmoved who see in it but one among other offsprings of a mystical habit of mind.

It is a difficult matter to judge how much repetition (in various verbal forms) may be necessary to make intelligible to others a mental standpoint with which one has become familiar, but which is probably strange to them. I do not indeed suppose that the theory of 'imaginary quantity' thus briefly sketched will at once be perfectly intelligible to every one whose interest in the subject has carried him on to this point; still less that, so far as it may be found intelligible, it will command ready assent. The vivid recollection of my own vacillating and blundering endeavours to attain freedom from mystical preconceptions forbids optimism in expectation. It will probably help to make clear this theory of 'imaginary quantity' and the standpoint from which it flows, if I bring them into contrast with the orthodox point of view and theory as set forth in Professor Chrystal's textbook. This will, in other words, be to show that the latter theory cannot be admitted by any one who admits the justice of the previous criticisms levelled against the orthodox exposition of elementary algebraic procedure, and that the former theory is not only consistent with, but inevitably flows from, these previous criticisms.

It is under the heading 'Digression on Imaginary and Complex Quantity',[1] and in connexion with the factorization of quadratic functions, that the validity of the new departure is discussed:

'Before proceeding to the factorization of a quadratic function in general, it is necessary to discuss briefly a fundamental point in the theory of Algebra which now arises for the first time. The special quadratic function x^2+c can, as has already been seen, be factorized by means of the identity $x^2-\lambda^2=(x-\lambda)(x+\lambda)$, provided always that c be a negative quantity, say $c=-d$, where d is an absolute arithmetic quantity. All we have to do is to determine λ so that $\lambda^2=d$ . . . In short, we may write $x^2+c\equiv x^2-(-c)\equiv(x-\sqrt{-c})(x+\sqrt{-c})$, so long as c is a negative quantity.

'If, however, c be a positive quantity, we can no doubt write $x^2+c=x^2-(-c)$; but the fundamental difficulty arises that we can no longer find a real quantity λ such that $\lambda^2=-c$. That this is so will be obvious when we reflect that the square of any quantity in the algebraic series of real quantity

$$-\infty, \ldots, -1, \ldots, 0, \ldots, +1, \ldots, +\infty$$

is positive.'

The argument contained in this last paragraph can be put in a form essentially equivalent to it, but simpler, so that we can more readily appraise its cogency. Suppose this positive quantity c to be equal to the positive quantity d^2. Then the argument would run, more simply and directly, thus: Although we may write $x^2+d^2=x^2-(-d^2)$, we cannot go on to write $x^2-(-d^2)=$

[1] *Introduction to Algebra*, pp. 184, 185.

$(x-\sqrt{-d^2})\ (x+\sqrt{-d^2})$, because $-d^2$ has no root, is not a product of identical factors, is not, in short, a power. In essence the argument amounts to this and to no more. But then it is natural to ask: If we may not write $\sqrt{-d^2}$ because this would be to imply that $(+d)\ (-d)$, or $-dd$, or $-d^2$ has a root, which is not the case, why are we permitted to write $-d^2$, which implies that $(+d)\ (-d)$ or $-dd$ is a power, which is also not the case? And if the answer to this question should happen to be that the expression $-d^2$ does not imply that we are dealing with an algebraic power, the inevitable rejoinder is: Why, then, should it be held that the expression $\sqrt{-d^2}$ implies that we are concerned with an algebraic root? We go on with the exposition:

'One way of meeting this difficulty would be simply to note and declare that the factorization of x^2+c by means of real operands is impossible when c is a positive quantity.

'There is, however, another course open to us. Although the laws of Algebra were derived from arithmetic, and we began by limiting the operands of Algebra to be arithmetical numbers, we have already passed beyond that limitation by introducing essentially negative quantity, the unit of which may be taken to be -1.'

Here again we must pause and protest. If there is anything really significant in the statement that we overpass the said limitation by introducing 'essentially negative' quantity, the statement is itself inadequate to the facts, for we must then also have overpassed this limitation by introducing 'essentially positive' quantity. I will not repeat here what I have already said on this point in the last chapter. But I will observe that the concept of quantity, simply, carries with it in my mind no implication of positivity or of negativity; so soon, however, as I associate this concept with that of opposition in the 'object of measurement'—be it money, force, or what not—then, by metaphor and for brevity's sake, I carry the opposition from the object to the measure. In algebra, as already remarked, this 'object' becomes merely quantity in the abstract. I resume the quotation from the point at which we last paused:

'Nothing hinders us from considering whether we might not still further enlarge the boundaries of Algebra by defining yet another kind of quantity having a new unit. The only point to be seen to is that any new kind of quantity must be such that we can operate with it together with the old kind of quantity by means of the laws and definitions of Algebra without landing ourselves in logical contradiction—in brief, without speaking or writing nonsense.

'Our immediate want is an algebraic quantity whose square shall be negative. Let us take the simplest case, and define a quantity i by the equation $i^2 = -1$. We call i the Imaginary Unit; and the understanding regarding it is that it is to be an algebraic operand; in other words, it is to be obedient to all the laws of Algebra. Whether it can be introduced without turning Algebra into nonsense will be seen by, and only by, operating with it and examining the consequences.'

With regard to the first of these two paragraphs, I would remark that we can speak or write nonsense without necessarily falling into logical contradiction. In so far, indeed, as speaking or writing nonsense is speaking or writing something which has no definite meaning, it is clear that the conditions for logical consistency or contradiction are simply absent. Next, it may be noticed that Professor Chrystal does not, as a matter of fact, assert that we can *conceive* another kind of quantity having a new unit. The proposition is that we shall enlarge the boundaries of Algebra by *defining* a new kind of quantity. Whether the word 'conceive' is here purposely avoided, or whether its absence is accidental; whether it is held to be of course implied that in defining a new kind of quantity we conceive it, or whether this implication is not intended—these are questions which naturally suggest themselves; but rational answers to them the orthodox standpoint seems unable to supply. Gauss, I believe, was the first to substitute the letter i for the combination $\sqrt{-1}$. If the meaning of this combination requires definition and we define it by means of the equation $\sqrt{-1} = i$, it is somewhat too obvious that a symbolic combination, meaningless until defined, is defined by another symbol also meaningless until defined. There is thus not a little art, and some humour, in defining i by the equation $i^2 = -1$, where the defining symbol has a definite meaning.

Being now in possession of the final step in the orthodox evolution of imaginary quantity, we can briefly review the entire argument, maintaining the two standpoints, the orthodox and the unorthodox, side by side, and comparing as we go the conceptual homogeneity of the two modes of viewing the development of elementary algebraic symbolism.

We find ourselves at the parting of the ways—the rational way and the mystical way—at the very outset of the science as now taught; and we take our first step along either the one path or the other according as the expression 'algebraic quantity' evokes in the mind a clear and precise, or a vague and mystical, conception. If we enter upon the rational way we shall begin by rejecting, as a confused assertion due to the reaction of algebraic symbolism upon the process of thought, the statement that in the first instance—as Cayley puts it—ordinary numbers or magnitudes are positive numbers or magnitudes.

The next point is the derivation of the conception of algebraic quantity. According to the orthodox view, this conception proceeds from the Law of Association for Addition and Subtraction, that is, the Law of Algebraic Summation, and in particular from the special cases $+(+a)=+a$, $+(-a)=-a$, $-(+a)=-a$, $-(-a)=+a$. Algebraic summation having apparently no intelligible meaning until algebraic quantity is conceived and its mode of symbolization explained, this view may, I think, be briefly described as putting the cart before the horse. According to the unorthodox view, algebraic or positive and negative quantity is a generalized conception, of which e.g. that of credit and debit is a particular instance. Starting with such symbols as $+a$, $-a$ for positive and negative quantity, the conventions which we adopt to express the transformation of simple algebraic sums into groups logically lead to the admission of additional symbolic forms for algebraic quantity, viz. $+(+a)$ and $-(-a)$ in addition to $+a$, $+(-a)$ and $-(+a)$ in addition to $-a$; and the law of association, or construction, which thus also becomes a rule of interpretation, is a statement of this fact. To derive the conception of algebraic quantity from these equivalent symbolisms for algebraic quantity seems to me to be the result of a confusion, of failing to keep clearly apart from one another the process of reasoning about quantity and the process of symbolizing the reasoning; and it is this confusion, still persistent, which is answerable for the subsequent unintentional sophistries propounded respecting the relation of magnitude between positive and negative quantity, sophistries which issue in the pseudo-concept of algebraic quantity as a series of quantity ascending in order of magnitude from $-\infty$ to $+\infty$.

In giving effect to the conventions governing the different forms of expressing an algebraic summation it becomes obvious to any one who does not allow the distinction between symbol and meaning to lapse from his mind, that in some of these forms the combinations $+a$, $-b$, &c., have already ceased unequivocally to symbolize algebraic quantities; they form part of organized expressions for algebraic quantity and have no definite meaning apart from the organized context in which they stand. Where in an algebraic sum we have algebraic quantities expressed as multiples or submultiples, transformations in accord with rule and convention issue in symbolic forms of which $(+a)$ $(-b)$, $-a/+b$ are instances. Since we know how, in the application of rule and convention, such forms arise, we can require no new rule for their interpretation, and we are thus also armed against the mystical illusions which they would otherwise suggest. But if we turn to the orthodox point of view we find that symbolic results of this type have been unwittingly allowed to distort the process of conception, and are presented as symbolic of the multiplication or division of one algebraic quantity by another, or of the positive or negative multiplication or division of a positive or negative quantity; and this pseudo-conception is explained (naturally in a highly artificial and unconvincing manner) as a sort of extension or modification of the ordinary conception of multiplication. It is, of course, quite permissible, from either standpoint, to say that in algebra the terms multiplication and division acquire an extension of meaning; but this would merely hide the real difference between the two views, which lies in the precise interpretation of this extension, in the answer to the question—What, exactly, is the nature of the conception corresponding to the expression 'algebraic multiplication', or, again, 'algebraic division'? According to the orthodox view, and confining ourselves for brevity's sake to multiplication, this conception appears to be that of a modification of, or generalization from (it is not clear which), the more simple arithmetical operation. According to what I venture to call the rational point of view there is no modification or generalization whatever of the *conception* of multiplication, or of division; these terms merely receive an *accretion* of meaning, that is, for the sake of brevity in exposition they are used to indicate at once the operation of multiplication or of division, and that of determining the sign of the quantity which results from either of the former operations.

If under the expression 'algebraic multiplication' we have allowed a pseudo-concept to gain possession of our minds, it will inevitably retain its hold upon us when we go on to algebraic involution and evolution. But here, to the mystical illusions produced by unclear apprehension of the relation between the two processes: the principal or conceptual and the subservient or symbolic, there is added a confusion of a quite different kind which springs from a discrepancy between the verbal phraseology of Algebra and the employment of corresponding specific symbols, an incongruity or logical inconsistency between the use of the terms 'root' and 'power' and the use of indices or exponents of power. We might almost at this point put the philosophy of the whole matter in a nutshell: To the latter (the logical inconsistency) is due the invention of 'imaginary quantity'; to the former (the mystical illusions) are due the obscurity of the discussions concerning the real import of this invention, and the difficulties which have stood in the way of recognizing the logical, as distinguished from the merely empirical sanction of its use.

On the other hand, if we reject this pseudo-concept of multiplication and division, we shall then be able to take the plain, straightforward, and rational view of involution and evolution which follows from a like view of multiplication and division. If we employ the expression 'algebraic factor' in the sense already suggested—that is, as denoting, not an algebraic quantity, but a member of the complex expression for an algebraic quantity—then we can also go on to use the term 'algebraic root' in a congruous sense, i.e. as the special name applied to an algebraic factor where all the factors involved in the complex expression are identical. If it is admissible to call an expression such as $(+a)(-b)(+c)\ldots$, or $\pm abc\ldots$, a product, it will be admissible to call an expression such as $(+a)(+a)(+a)\ldots n$ terms, briefly written $(+a)^n$, a power, the factor $(+a)$ being called a root. Is there anything to hinder us, if we find it convenient, from extending this use of the power exponent to cases where the factors differ only in sign, and where the expression which they constitute cannot, in accordance with the rule of signs, be replaced by another in which the signs are all alike? Only the bare fact that this extension involves a formal inconsistency in symbolism, for it does not induce us to false conclusions.

But if, for the sake of its convenience, we do admit this extended application of the power index, it is a consequence of this admission that we can further, and logically, admit an analogical extension of the use of the root index or radical sign, so that it may be said that the invention of 'imaginary quantity' rehabilitates the symbolic system, formally re-establishes its logical consistency. For if this artifice enables us to do what we could not do before, viz. express certain algebraic products as products of identical factors, then these factors and products will have precisely the same claim to the designations 'root' and 'power' as those other factors and products which have already had these names assigned to them. The only consideration which makes acquiescence in this artifice difficult, notwithstanding its logical filiation with those which have gone before, is one which springs from looking at algebraic symbolism from a mystical instead of from a rational standpoint. When it is proposed to equate $(+a)(-a)$ to $(a\sqrt{-1})(a\sqrt{-1})$, the 'difficulty' is felt that, while in the first of these expressions the factors $+a$, $-a$ respectively symbolize a positive and a negative quantity, in the second the factor $a\sqrt{-1}$ is uninterpretable as a quantity. But it is just this mystical attitude of mind, itself the result of a damaging reaction of the symbolism of algebra upon the individual's judgement of his own conceptual process, which leads the algebraist to give this interpretation to the factors in the expression $(+a)(-a)$. That 'the symbol $(-1)^{\frac{1}{2}}$ is absolutely without meaning when it is endeavoured to interpret it as a number' is of course merely another way of stating the difficulty, but it would never have been felt by the algebraist had not the tendency to mysticism overpowered his judgement; had it been possible for him to withstand this tendency he would never have regarded this symbol otherwise than as a constituent part of an expression which does symbolize an algebraic number or quantity, viz. $(-1)^{\frac{1}{2}} \times (-1)^{\frac{1}{2}}$; and we should probably never have heard of conceiving or defining a new kind of quantity whose unit is $\sqrt{-1}$ or i.

Number and Quantity are neither arithmetical nor algebraic, but Arithmetic and Algebra are technical languages designed to facilitate reasoning about number and quantity. Quantity, as such, is neither positive nor negative; the relation of opposition expressed by the correlative terms positive and negative is not, as such, quantitative; but quantitative relation and this relation of opposition can be jointly predicated of the things which we apprehend under these conjoined aspects. 'Algebraic' number or quantity denotes, in my mind, the synthesis in apprehension of these two relations; but the relations do not modify one another in the synthesis: the apprehension of opposition in quantitative comparison no more constitutes an extension or modification of the abstract idea of quantity than the quantitative aspect of opposition modifies or extends the latter abstraction. When we combine the abstract idea of direction with that of quantity under the term vector or directed quantity, is it urged that we thus conceive a new kind of quantity, a new kind of direction? The expression 'algebraic quantity', useful as an abbreviation for 'positive and negative quantity'—and in the latter expression we are already metaphorical in our desire to be brief—deceives us if we forget, just on those occasions when we ought to remember, that we speak in figure. It is evident, from the manner in which algebraists express themselves in explaining the principles of their symbolism, that they sometimes confuse, where it is important to distinguish, between symbol and concept. In no other way, at least, does it seem possible to account for the succession of pseudo-concepts which we have found in the received exposition of the principles of the symbolism.

To the algebraists of an earlier day, contemplating symbolic forms which seemed to suggest, without actually denoting, some new conception of quantity, it may very well have appeared that the process of symbolization had, in some unexplained, mysterious way, outrun the process of definitely conscious conception. At the same time, the connexion between geometry and the calculus, especially the influence of geometrical conceptions on the development of the calculus, obvious in the use of such terms as 'square' and 'cube' to denote certain numerical relations, could not but influence the speculations of mathematicians. In the early years of the nineteenth century the idea was 'in the air', circumambient, of connecting these enigmatic algebraic forms, the symbols of so-called imaginary and complex number or quantity, with geometrical conceptions;

of finding in geometrical relations an interpretation of this development of algebraic symbolism. The success of this endeavour appeared to many, and not least to the innovators themselves, as the solution of the problem, the purely mystical problem, which had so long perplexed mathematicians, and naturally enough led to the paradoxical conclusion that Algebra was and always had been dependent upon Geometry for its development. In so far as Argand and other mathematicians devised a systematic method of employing those algebraic expressions which involve the so-called symbols of imaginary quantity, to denote certain geometrical relations, Algebra may be said to have ceased to be simply, as Newton called it, Arithmetica Universalis, because the relations thus symbolized are not relevant—though given *names* of relations which *are* relevant—to number or quantity in the abstract. But, apart from this special application of algebraic symbolism, Algebra remains Arithmetica Universalis, because these so-called symbols of imaginary quantity are then merely artificial factors in the expression of relations relevant to number or quantity in the abstract. This distinction (which illuminates anew the difference between Mr. Whitehead's standpoint and my own) will, I suppose, be perfectly intelligible to every one who has with any care analysed what is known as 'the diagrammatic method of representing complex numbers'; but as the general reader is probably unacquainted with the subject, or may have forgotten what he once knew of it, I will say something here about the ideas in which the method originated.

Argand's *Essai sur une Manière de représenter les Quantités imaginaires dans les Constructions géométriques* was published in 1806, and seems at the time to have passed almost unnoticed. It was brought more or less prominently to the notice of mathematicians in a roundabout way. Argand had submitted his essay to Legendre, Legendre had given a brief account of it in a letter to the brother of J. F. Français, professor in the imperial school of artillery and engineers. Français found this letter among his deceased brother's papers, and, basing himself upon its contents, wrote a paper entitled 'Géométrie de Position', which he published in Gergonne's *Annales des Mathématiques*,[1] at

[1] Tome iv, pp. 61–71.

the same time explaining that the fundamental ideas were not his own, and expressing the wish that the original author of them would make known his work. This drew from Argand an account, published in the same volume of the *Annales*,[1] of his Essai.

Argand introduces the subject with some remarks on real and imaginary quantity. According to him, and briefly put, the subtraction of a quantity from a less quantity being impossible, the operation is imaginary or the result is an imaginary quantity; but the idea of such a quantity as real may be deduced thus: If a is the resultant weight acting on the scale A of a balance with weights in both scales, then since it is possible to diminish a either by taking weights from the scale A, or by adding weights to the scale B, the series . . . $+3a$, $+2a$, $+a$ can be prolonged beyond zero, and $-a$, $-2a$, $-3a$ . . . will be quantities as real as $+a$, $+2a$, $+3a$ . . .

From this we see that the idea, dominant to this day, that quantity is originally 'positive', and that quantity as 'negative' is a subsequent generalization or deduction, is at least a hundred years old. Argand does not tell us why, if the measure of the resultant weight is to be symbolized by a, it is subsequently symbolized by $+a$. Of course we see why: because the weight in the one scale is considered, and can only intelligibly be considered, as positive (or negative), in contradistinction with the weight in the opposite scale. But then the deduction is superfluous, merely leads back to the conception from which it starts; and we are surely in no need of argument to convince us that a difference of weight is as real when the excess is in one scale as when it is in the other.

The fundamental notion of the Essai is to use the symbol $\sqrt{-1}$ to express the relation of perpendicularity. That is, 1 and $\sqrt{-1}$ are to denote the same unit of length measured respectively on two straight lines perpendicular to one another, opposition in the sense of measurement being symbolized as usual by the signs + and − placed before the symbol of quantity. This is also the case with the 'Mémoire sur les Quantités imaginaires' of the Abbé Buée, communicated to the Royal Society in 1805, and printed in the 1806 volume of the *Philosophical Transactions*. Argand, Français, and Buée all supposed that the attribution

[1] pp. 133–47.

of this meaning to the symbol $\sqrt{-1}$ required, and was susceptible of, rigorous demonstration. But this was a misapprehension of the nature of the mental process which led them to propose this convention, and the effect of it was to retard the recognition of the new doctrine, for not unnaturally those mathematicians who expected to find what, in the nature of the case, could not be given, were blind to its intrinsic merits.[1]

Argand himself seems to have been among the first to realize that 'probably' no purely logical demonstration can be given of this interpretation; he even went so far as to show[2] that from one point of view (which is, indeed, the right point of view) the quest of such a demonstration is an irrelevance. The explanation of these attempts deductively to establish the fundamental idea of the new theory lies, it seems to me, in this: that its inventors looked upon it, or were strongly inclined to look upon it, as a solution of the enigma of imaginary quantity in Algebra, and not simply as an ingenious adaptation of the symbol $\sqrt{-1}$ (by itself meaningless in algebra), and of the algebraic forms in which it is involved, to the expression of certain geometrical relations. And, clearly, this is a point of view from which the mystical conclusion would readily follow that Algebra is not, and never could have been, merely Arithmetica Universalis, but in its very essence is, and always must have been, a symbolic system deriving from geometrical conceptions.

Having shown how an imaginary quantity in arithmetic is physically interpretable ('Cette distinction des grandeurs en réelles et imaginaires est plutôt physique qu'analytique'), Argand proceeds by analogy to do the same for imaginary quantity in algebra:

'When in geometry (sous le point de vue appelé rapport géométrique) we compare two quantities of a kind susceptible of yielding negative values, the idea of this comparison is evidently complex. It consists (1) of the idea of numerical relation, or relation of absolute magnitude; (2) of the idea of the relation of directions or senses to which the quantities belong: a relation which, in this case, can only be that of identity or of opposition. Thus, when we say that $+a : -b :: -ma : +mb$, we mean not only that $a : b :: ma : mb$, but we also affirm that the direction of

[1] See the criticism of it by Servois in the *Annales*, vol. iv, pp. 228 et seq. Gergonne, who comments on this criticism, is more clear-sighted.

[2] *Annales*, tome v, pp. 197 et seq.

the quantity $+a$ is, relatively to the direction of the quantity $-b$, what the direction of $-ma$ is relatively to the direction of $+mb$; and we may even express this last conception in an absolute manner by writing,

$$(A) \quad . \quad . \quad . \quad . \quad +1 : -1 :: -1 : +1.$$

'Be it now proposed to determine the mean proportional between $+1$ and -1, that is, to determine the value of x which satisfies the proportion $+1 : x :: x : -1$. Since x cannot be equated to any number either positive or negative, it seems we must conclude that the quantity sought is imaginary.

'But since we have already found that negative quantities, which appeared at first only to exist in imagination, acquire a real existence when we combine the idea of absolute magnitude with that of direction, analogy prompts us to inquire whether we cannot reach a similar result with respect to the quantity x.

'Now if there is a direction d such that the positive direction is to d as d is to the negative direction, then, denoting by $1d$ the unit of length taken in the direction d, the proportion

$$(B) \quad . \quad . \quad . \quad . \quad . \quad +1 : 1d :: 1d : -1$$

will present (1) a purely numerical proportion $1 : 1 :: 1 : 1$, (2) a proportion or similitude of relations of direction analogous to that of the proportion (A).'

The rest of the argument, so far as the fundamental notion is concerned, may briefly be summarized thus: If from the point O we take a unit of length in the directions OA, OB, and OC perpendicular to AB, the conditions for realizing the proportion (B) are fulfilled. OC is a mean proportional between $+1$ and -1; and since $\sqrt{-1}$ is the algebraic mean proportional between $+1$ and -1, we can write $OC = \sqrt{-1}$.

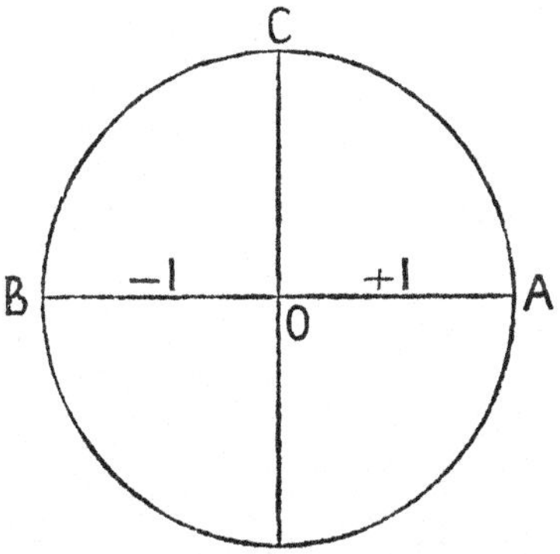

This argument expresses a process of thought in which there is more than mere deduction from admitted premisses. Its essential trait is origination or invention—a creation of new premisses, but put into a form which tends to obscure its real nature. That a similitude or identity of relations of direction, such as that to which

Argand draws attention, is a 'proportion', and that a direction perpendicular to two opposed directions is a 'mean proportional' between them, is a proposition which is neither true nor false, and thus cannot be made the subject of proof or disproof. But, given the purpose and the object of Argand's Essai, it is legitimate to propose, as a new premise or convention in symbolism, that identity of relations of direction shall be given the same name, and shall be algebraically symbolized in the same way, as identity of relations of magnitude. Any one who, reading Argand's argument, looks upon this proposition as one which ought to be demonstrated, will naturally find in the conclusion that $\sqrt{-1}$ denotes the relation of perpendicularity nothing but the reassertion of an unproved proposition. But from the moment this proposition is understood and admitted as a new convention in symbolism, it is at once seen that the employment of $\sqrt{-1}$ to denote this geometrical relation follows logically from this new premise in symbolism. For the equation $\sqrt{-1} \times \sqrt{-1} = -1$ can, in accord with algebraic convention, be transformed into $+1 : \sqrt{-1} :: \sqrt{-1} : -1$, and since the latter expression does *not* symbolize a quantitative proportion ($\sqrt{-1}$ not being a symbol of quantity), $\sqrt{-1}$ can without contradiction or ambiguity be substituted for $1d$ in equation (B). Again, $\sqrt{-1}$ or i not symbolizing a quantity, the combination $a+bi$ does not symbolize the conception of *summation*, and is, thus isolated, meaningless, being nothing but a factor in an organized expression of algebraic quantity; it can thus be employed without ambiguity or contradiction to symbolize the geometrical *combination* of $OA = a$ and $AB = bi$, or the resultant, OB, of this combination: in the ordinary language of Algebra, OB *represents* the complex number $a+bi$.

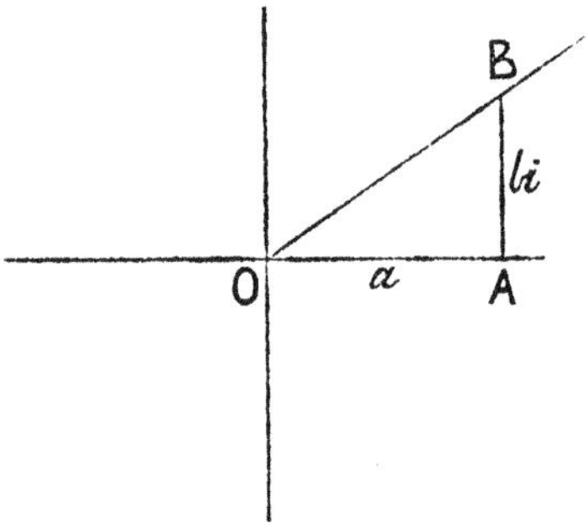

I said a little while ago that in so far as these geometrical conceptions are assigned as meanings of the so-called symbols of imaginary and complex quantity or number, Algebra may be said to be no longer simply Arithmetica Universalis. The same may be said with reference to the totally different interpretation of these forms in analytical geometry. But I should express my meaning more exactly in saying that in these cases Algebra becomes something more than, without ceasing to be, Arithmetica Universalis. For algebraic symbolism is not an expression of the development of geometrical conception; it is and it remains—notwithstanding these particular attributions of geometrical meaning—essentially and directly a language specially devised for the purpose of reasoning about number or quantity in the abstract; i.e. it is and remains what Newton called it, Arithmetica Universalis. And in order to make my meaning still more plain, if it is not already plain enough, I will add that in saying that Algebra is essentially Arithmetica Universalis, I intend to traverse the assertion that Algebra is independent of Arithmetic, unless nothing more is meant by this than to reiterate in other words the perfectly obvious and indisputable fact that the algebraic forms in question are arithmetically uninterpretable. But it is clear there is more than this in the assertion; the independence is looked upon as an inference or conclusion arising from other considerations, one of which is this particular case of uninterpretability in arithmetic: 'If there were such a dependence, it is obvious that as soon as algebraic expressions are arithmetically unintelligible all laws respecting them must lose their validity.'

But the algebraic expression $\sqrt{-1} \times \sqrt{-1}$ *is* arithmetically intelligible in the sense demanded by this passage. The point is whether it is legitimate to isolate $\sqrt{-1}$ and regard it as an algebraic 'expression' or entertain such a question as, What does it mean? No doubt it will appear to be so if we submissively follow the orthodox exposition of the development of elementary algebra, and are either blind, or content to shut our eyes, to the sophisms and incongruities which abound in it. The argument of this and the preceding chapter is a sustained protest against the mystical bias which prompts these sophisms and incongruities and renders them not intolerable even to critical minds, against the cumulative and finally disastrous effect of an unsuspected reaction of the symbolic forms of algebra upon the conceptual process which gives them birth. Set free the judgement from this mystical bias, disengage the process of conception from the false implications of the process of symbolism, and not only do these sophisms and incongruities disappear, but with their disappearance also vanishes the suggestion which they facilitate, viz. that such combinations as $\sqrt{-1}$, $a+bi$ are, in and by themselves, algebraic 'expressions', that is, symbolic of mathematical notions or abstractions. They are then at once recognized as mere factors or constituent parts of actual expressions, algebraically symbolic, interpretable in terms of generalized number and abstract quantity.

CHAPTER IX

THE DOCTRINE OF IMAGINARY LOCI IN GEOMETRY

Explanation of Imaginary Points (1) by means of the Principle of Contingent Relations (Chasles), (2) by means of the Theory of Involution (von Staudt).—The Doctrine of Geometrical Imaginaries, rationally considered, is an artifice in expression which involves paradox for the sake of brevity in the statement of certain geometrical relations.—The conceptions of the point and line at infinity as leading to this paradoxical mode of expression.—Algebraic Imaginaries and Analytical Geometry.

THE common sense of this doctrine—using the term 'common sense' in place of the more pretentious word 'philosophy'—as Clifford did in the *The Common Sense of the Exact Sciences*—lies, so it appears to me, in a rational interpretation of, e.g. the simple statement which Cayley makes on the subject in his presidential address: 'But the geometrical construction being in each case the same, we say that in the second case also the line passes through the two intersections of the circle.' This, as will be recollected, refers to the configuration composed of two circles and a straight line, and the two cases, (1) where the line is the common chord of the two circles, (2) where the line is not a chord at all.

This explanation of Cayley's, apparently not intended by him as anything more than a bare statement of the practice of geometers, really contains, in epitome and by reference to a particular case, all that Chasles has to say on the subject in the fifth chapter and note xxvi of the *Aperçu historique sur l'Origine des Méthodes en Géométrie*; that is, his statement and discussion of the *Méthode ou principe des relations contingentes*, of the distinction between the accidental and permanent properties of a system of figures. In immediate connexion with the latter phrase, he discusses (p. 205) the case of the system of two circles and line in a plane which Cayley used in illustration of the doctrine. He says:

'When the two circles intersect, this straight line is their common chord, and this fact suffices to define and to construct it: this is what we call one of its contingent or accidental properties. But when the two circles do not intersect, this property disappears although the line still exists and is of great utility in the theory of circles. We must therefore define and construct this straight line by means of some other one of its properties common to all cases of the general construction of the figure, or of the system of the two circles. This will be one of its permanent properties. It is through such considerations that M. Gaultier, instead of calling this line the common chord of the two circles, is led to call it the *radical axis* ; an expression prompted by a permanent property of this line, viz. that the tangents drawn from any one of its points to the two circles are equal, so that every point of this line is the centre of a circle which cuts the other two orthogonally. . . .

'The doctrine of contingent relations seems to us to present a further advantage, that of affording a satisfactory explanation of the word "imaginary", now used in pure geometry, where it expresses *un être de raison sans existence*, but which we can nevertheless endow with certain properties of which we make use temporarily as auxiliaries, and about which we reason as we do about a real and palpable object. This idea of the imaginary, which seems at first sight obscure and paradoxical, thus acquires, in the theory of contingent relations, a clear, precise, and legitimate sense.'

Chasles here refers us to note xxvi, in which he enters more fully into his view of the doctrine of contingent relations as explanatory of the use of imaginaries in pure geometry. We employ this word 'imaginary', he says, 'as a means of attaining brevity of expression ; it implies that the process of reasoning applies to another general state of the figure, in which the parts which are the subject of reasoning would really exist, instead of being, as in the figure actually contemplated, imaginary. And since according to the principle of contingent relations, or, if you prefer it, the principle of continuity, the truths demonstrated for one of the two general states of the figure apply to the other state, it is seen that the use of imaginaries is completely justified.' In other words, and with reference to the system of two circles, when we reason about the general case in which the two circles do not intersect, the reasoning applies also to the general case in which the two circles do intersect, because it is directed to some relation between the two circles and the line, which is independent of the accidental circumstance of intersection, or of non-intersection. Thus when we say that the radical axis of two circles is a line which passes through the points, real or imaginary, of intersection of the two circles, this is a brief and, let us add, a paradoxical way of saying that the radical axis of the two circles is a line which has a relation to the two circles which is independent of the contingency of intersection or non-intersection.

This is a common-sense or rational explanation of the use of a paradoxical mode of expression which, *inter alia*, absolves us from attending to such ambiguous if not mystical definitions as that an imaginary geometrical element is *un être de raison sans existence.*[1] The term 'imaginary', used in connexion with the names of geometrical entities, is thus, according to Chasles, a means of giving expression to the geometrical facts involved in what Poncelet calls 'the principle of continuity' (*Traité des Propriétés projectives des Figures*) or, as Chasles himself calls it, 'the principle of contingent relations'. This principle is purely geometrical: in Cayley's words, we arrive geometrically at the notion of imaginary *loci*. Later, von Staudt also shows how, through purely geometrical considerations, we are led to introduce imaginary elements into geometrical reasoning; but the principle which underlies his explanation is, in essence, the same as Chasles's: that of distinguishing between accidental or contingent and permanent relations, the term 'imaginary' and also the term 'real' being relevant to the contingent relations.[2]

But in von Staudt's exposition, side by side with the rational explanation of the use of paradoxical expressions, are there not also mystical propositions, i.e. propositions left without interpretation in the language of common sense, thus again prompting such questions as: What is an imaginary point? what are imaginary geometrical elements? what is imaginary space, or the *locus in quo* of imaginary points, lines, and figures? —questions which, if we wittingly employ paradoxical modes of expressing definite geometrical relations, could not be asked at

[1] *Un être de raison* is an expression common enough with French philosophical writers. In one sense of the word 'existence', *un être de raison* has no existence; in another sense of this word, the existence of *un être de raison* consists in our conceiving it. *Un être de raison sans existence* is thus a highly ambiguous piece of phraseology.

[2] We require the term 'real' in this connexion only because we use the term 'imaginary'. If we express ourselves literally, we want neither the one nor the other.

all, because they would be recognized as quite meaningless. If such questions are still asked—either *sub silentio*, or overtly, as when Cayley asks, What is an imaginary point?—is this not plain evidence of continued subjection to a mystical tendency? The point here is not that the tendency to the mystical definitely issues in an unquestioned belief that these paradoxical expressions are expressions of hitherto unconceived geometrical entities (which would be the case if these expressions were interpretable literally); such a belief is, in *pure* geometry, continually held in check or thwarted by the mere fact that the process of reasoning in this science necessarily involves the continual presence of representative or typical imagery. The tendency to the mystical has not here the free play which it has in algebra, and the effect of this difference is very clearly marked in Cayley's presidential address: he does not ask, with reference to imaginary quantity in algebra, what he asks with respect to imaginary elements in geometry, although he invites philosophical discussion of the doctrine as a whole. It seems to have been for him beyond question that the phraseology and symbolism of imaginary quantity express a real development of the conception of abstract quantity, that in algebra we attain, and express by this phraseology and symbolism, the notion of a new kind of abstract quantity, just as, at an earlier period of development, we arrived at the notion of 'essentially negative' abstract quantity.

I return to the question suggested above with regard to von Staudt's exposition. There is an abridgement of this given by Professor Henrici in the article 'Projection', *Ency. Brit.* vol. xix, p. 799. I will reproduce so much of it here as, with a running commentary, will suffice to show clearly how the question is prompted:

'If a line cuts a curve and if the line be moved, turned for instance about a point in it, it may happen that two of the points of intersection approach each other until they coincide. The line then becomes a tangent. If the line is still further moved in the same manner it separates from the curve and two points of intersection are lost. Thus in considering the relation of a line to a conic we have to distinguish three cases—the line cuts the conic in two points, touches it, or has no point in common with it. This is quite analogous to the fact that a quadratic equation with one unknown quantity has either two, one, or no roots. But in algebra it has long been found convenient to express this differently by saying a quadratic equation has always two roots, but these may be either both real and different, or equal, or they may be imaginary. In geometry a similar mode of expressing the fact above stated is not less convenient.

'We say, therefore, a line has always two points in common with a conic, but these are either distinct, or coincident, or invisible.'

The writer, following Clifford, uses 'invisible' throughout for 'imaginary'. So far the paradoxical expressions are defined in ordinary language.

'Invisible points occur in pairs of conjugate points, for a line loses always two visible points of intersection with a curve simultaneously. This is analogous to the fact that an algebraical equation with real coefficients has imaginary roots in pairs. *Only one real line can be drawn through an invisible point,* for two real lines meet in a real or visible point. *The real line through an invisible point contains also its conjugate.*

'Similarly there are invisible lines—tangents, for instance, from a point within a conic—which occur in pairs of conjugates, two conjugates having a real point in common.'

Here we have a number of paradoxical propositions so far not given a rational interpretation. It is, possibly, the recognition of this fact which prompts the following observations:

'The introduction of invisible points would be nothing but a play upon words unless there is a real geometrical property indicated which can be used in geometrical construction—that it has a definite meaning, for instance, to say that two conics cut a line in the same two invisible points, or that we can draw one conic through three real points and the two invisible ones which another conic has in common with a line that does not actually cut it. We have in fact to give a geometrical definition of invisible points. This is done by aid of the theory of involution.'

Clearly no real geometrical properties have so far been assigned or indicated by the paradoxical expressions of the two previous paragraphs.

'An involution of points on a line has either two or one or no foci. Instead of this we now say it has always two foci which may be distinct, coincident, or invisible. These foci are determined by the involution, but they also determine the involution. If the foci are real this follows from the fact that conjugate points are harmonic conjugates with regard to the foci. That it is also the case for invisible foci will presently appear. If we take this for granted we may replace a pair of real, coincident, or invisible points by the involution of which they are the foci.'

That is to say: take it for granted, first, that a definite meaning will be assigned to the statement that invisible foci are *determined* by an involution which has no real ones, and vice versa, and next that a definite meaning will be assigned to the statement that conjugate points are harmonic conjugates with regard to invisible foci; then you may take it for granted that a definite meaning will be assigned to the statement that a pair of invisible points may be replaced by the involution of which they are the foci. Rather a perplexing mode of leading up to what is subsequently found to be a very simple matter.

'Now any two pairs of conjugate points determine an involution. *Hence any point-pair, whether real or invisible, is completely determined by any two pairs of conjugate points of the involution which has the given point-pair as foci and may therefore be replaced by them.*'

Read: Hence you may take it for granted that a definite meaning will be assigned to the statement that every invisible point-pair is completely determined by, &c.

'Two pairs of invisible points are thus said to be identical if, and only if, they are the foci of the same involution.'

Finally we come to what is perfectly intelligible independently of whether we have or have not assigned definite meanings to the foregoing paradoxical expressions.

'We know that a conic determines on every line an involution in which conjugate points are conjugate poles with regard to the conic—that is, that either lies on the polar of the other. This holds whether the line cuts the conic or not. Furthermore, in the former case the points common to the line and the conic are the foci of the involution. Hence we now say that this is always the case, and that the *invisible* points common to a line and the conic are the *invisible* foci of the involution in question. If then we state the problem of drawing a conic which passes through two points given as the intersection of a conic and a line as that of drawing a conic which determines a given involution on the line, we have it in a form in which it is independent of the accidental circumstance of the intersections being real or invisible. So also is the solution of the problem. . . .'

This explanation is quite clear. We may also, I presume, put it a little differently, thus:

1. A conic determines on any line, whether the conic cuts the line or not, an involution in which, &c.

2. If the line cuts the conic, the points of intersection are the foci of the involution; if the line does not cut or touch the conic, the involution has no foci.

3. In the latter case, we say that the line cuts the conic in invisible points, or in the invisible foci of the involution.

4. In accordance with this paradoxical mode of expression, the problem: To draw a conic which determines on a line a given involution, may be expressed otherwise as: To draw a conic which passes through two points given as the intersection, real or invisible, of a conic and a line. Similarly the theorem: Through three given points we can always draw one conic, and one only, which determines on a given line a given involution, may also be expressed thus: It is always possible to draw a conic which passes through three real given points and through two invisible points which any other conic has in common with a line.

The non-mathematical reader may ask, What is the use of giving paradoxical instead of literal expression to geometrical relations? This question is not very easy to answer in a few words, nor without reference to the expressions: the point, the line, the plane 'at infinity'—themselves paradoxical, or semi-paradoxical, expressions of geometrical conceptions. But I think it may be said briefly that the gain by paradox in expression is that in many cases it enables the geometer to state, or to reason about, geometrical relations in language which is far more concise than that of literal expression. Rationally understood, paradox in geometrical expression is a labour-saving device.

But is it always rationally understood or considered? As we have just seen, the paradoxical expressions of the above problem and theorem are explicable without reference to the previous paradoxical expressions which are supposed to lead up to them and explain them. In so far as these previous paradoxical statements are translatable into language which, taken literally, express real geometrical properties, they are so only in virtue of the equivalence of literal and paradoxical expression subsequently given in relation to the problem or theorem. Given this equivalence, we can see how these antecedent expressions originate. Thus 'invisible points occur in pairs of conjugate points'. The explanatory addendum, 'for a line loses always two visible points of intersection with a curve simultaneously' is clearly no explanation whatever until we know that the geometrical relation between a line and a conic, called the involution determined by the conic on the line, is to be paradoxically expressed, and how; and when we know this the addendum becomes superfluous. But do all these prior statements express real geometrical properties? For instance, this, which, in the text, is italicized: '*Only one real line can be drawn through an invisible point*, for two real lines meet in a real or visible point.' We are reminded here of the subsequent phrase: 'we have in fact to give a geometrical definition of invisible points.' But the definition given in the statement of the problem, To draw a conic, &c., is not a definition of invisible points, it is the definition of a phrase in which the term 'invisible point' occurs. Apart from the context in which it there stands, taken in isolation, this term has no definite meaning. The case is analogous with that of 'imaginary quantity', or $\sqrt{-a}$, which, isolated from the *purely algebraic* expressions in which it occurs, has no definite meaning. We might almost say that the term 'invisible point', in geometry, like the symbol $\sqrt{-a}$ in algebra, is merely a 'factor' in paradoxical expressions of real meaning.

Now does this statement, only one real line can be drawn through an invisible point, express a real geometrical relation or property, distinct from that subsequently defined both literally and paradoxically in the problem of drawing a conic, &c.; or is it prompted by the mystical belief that the words 'an invisible point' simply, in isolation, *must* have a definite meaning? In the former case, what is this geometrical property or relation, literally expressed? I do not know, and the writer does not tell us. In the latter case, from the premisses that an invisible point is some kind of point but not a real point, logic would force the conclusion that only one real line can be drawn through an invisible point, and, mystically again, this conclusion would be admitted as expressing a real geometrical property or relation. Now let it be supposed that the expression 'an imaginary point' must have a definite meaning, and nothing is more certain than that the question, What is an imaginary point, will present itself, and present itself as a rational question.

The doctrine of geometrical imaginaries, then, freed from all mystical implication, is, to speak plainly, not a geometrical doctrine at all, not a new conception of geometrical properties or relations, but a new and paradoxical mode of expressing geometrical properties or relations, a labour-saving device which, in skilled hands, is very effective. But it is not surprising that so free a use of paradox in language should at the same time give rise to mystical illusions, and not least in the minds of those who use it with consummate skill in geometrical investigation. And, if I am not mistaken, lapse into mystical illusion is here (as in algebra) rendered the easier, the more inevitable, from the nature of the anterior development of the science, of the mode of geometrical reasoning which Monge was the first to reduce to a method. In the first principles of this method are involved propositions which, in their expression, verge upon the paradoxical, but which may be accepted literally by the learner without too great a shock to his sense of the rational, and thus insensibly incline him to subsequent mystical interpretations. I refer to what are called the notions of the point, line, and plane at infinity.

How are these notions introduced and explained? Let us take the account given of them in the article on projective geometry, vol. x of the *Encyclopaedia Britannica*. We find there, at the outset, that a number of Euclid's preliminary propositions, taken from the Definitions, Postulates, and Axioms of the first book of the *Elements*, are grouped together and expressed anew in the following way:

I. A plane is determined—	A point is determined—
1. By three points which do not lie in a line;	1. By three planes which do not pass through a line;
2. By two intersecting lines;	2. By two intersecting lines;
3. By a line and a point which does not lie in it.	3. By a line and a plane which does not pass through it.
II. A line is determined—	
1. By two points.	2. By two planes.

Attention is then called to the fact that some of these propositions are not strictly true without the addition of some such qualifying words as 'if they are not parallel'. In order 'to correct this we have to reconsider the theory of parallels'.

The writer then proceeds to this reconsideration; but before we follow him, let us note that in the introductory remarks it is explained that the characteristic, or one of the characteristics, of the modern method as compared with the old is a striving towards greater generality. Thus, on the modern view 'a straight line is considered as given in its entirety, extending both ways to infinity, while Euclid is very careful never to admit anything but finite quantities. The treatment of the infinite is in fact another fundamental difference between the two methods. Euclid avoids it. In modern geometry it is systematically introduced, for only thus is generality obtained.' I quote this because it bears upon the subsequent reconsideration of the theory of parallels, also because I should like to draw attention to what appears to me the ambiguity of the statement that according to the modern method a straight line is considered as given in its entirety, extending both ways to infinity. If by an infinite straight line we understand a length which increases without limit, it is not clear how we can consider it as given in its entirety; this expression seems more appropriate to a line given as finite. A whole is finite, an infinite whole is an expression which lends itself too easily to the criticism that it is prompted by a confusion of thought.

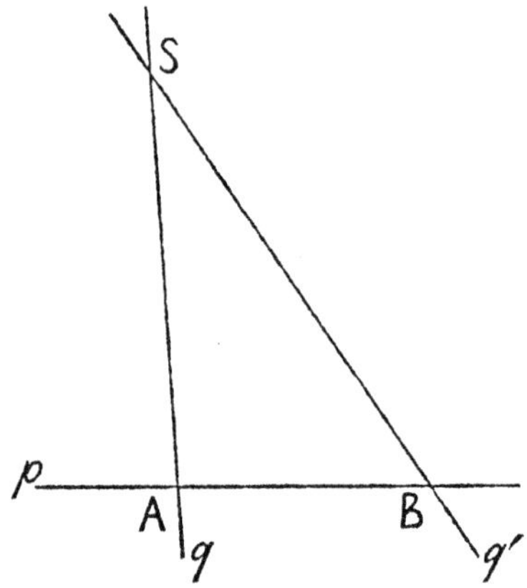

Now for the reconsideration of the theory of parallels:

'Let us take in a plane a line p, a point S not in this line and a line q drawn through S. Then this line q will meet the line p in a point A. If we turn the line q about S towards q', its point of intersection through p will move along p towards B, passing, on continued turning, to a greater and greater distance, until it is moved out of our reach. If we turn q still farther, its continuation will meet p, but now on the other side of A. The point of intersection has disappeared to the right and reappeared to the left. There is one intermediate position where q is parallel to p—that is, where it does not cut p. In every other position it cuts p in some finite point. If, on the other hand, we move the point A to an infinite distance in p, then the line q which passes through A will be a line which does not cut p at any finite point. Thus we are led to say: *Every* line through S which joins it to any point at an infinite distance in p is parallel to p. But by Euclid's 12th axiom there is but one line parallel to p through S. The difficulty in which we are thus involved is due to the fact that we try to reason about infinity as if we, with our finite capabilities, could comprehend the infinite. To overcome this difficulty, we may say that all points at infinity in a line *appear* to us as one, and may be replaced by a single "ideal" point.' This point is called 'the point at infinity' in the line.

The reference to our finite capabilities is a salutary reminder, though we are perhaps never less in need of it than when our attention is claimed by questions which touch upon the foundations of knowledge and reasoning. With our finite capabilities, however, we either have or have not found a pair of contradistinct concepts to which we give these names 'finite' and 'infinite'; and, plainly, if we *do* attribute definite meanings to these terms, these meanings are at once mutually exclusive and correlative. If we do not, then the reasoning in which they are involved becomes a mere play upon words. The fact is that the finitude of our capabilities is here an irrelevant consideration. Every one admits that we here form a general conception of lines as finite, but not every one pauses to reflect upon the process of doing it. So soon as he does, he cannot but become aware that he only thinks of lines *as finite* (or *as infinite*) by contradistinction with the thinking of them *as infinite* (or *as finite*); literally, it may be said, as finishing here or there, and as unfinishing, finishing nowhere, extending without limit. But then he will also recognize that such expressions as 'extending both ways *to* infinity', '*at* an infinite distance', are metaphors (not to call them paradoxes) convenient only if we do not let them delude us. Now are we not deluded if we accept the above deduction of the 'point at infinity' in the line, instead of seeing in this phrase nothing but a convenient way of escaping from the exception in the theory of projection and correspondence which necessarily arises from the admission of Euclid's theory of parallels?

Look back to the argument. The writer makes it appear that by first considering the rotation of a line *q* and the concurrent indefinite recession and reappearance of the point of intersection with *p*, and then next considering the indefinite recession of the point of intersection and the concurrent rotation of the line, there arises in connexion with Euclid's 12th axiom a sort of contradiction or incongruity which is to be attributed to our inability to comprehend the infinite, to our finite capacities. But if we examine the argument carefully I think we shall see that there is no need to invoke the finitude of our capacities in order to explain the contradiction or incongruity which manifests itself in the reasoning, and that the difficulty in which we find ourselves simply arises from an uncritical use of the instrument of reason, viz. language. We begin by substituting, for a cumbrous but literal description of what we conceive under the term 'infinity' in connexion with a line, a concise but metaphorical or figurative expression, and the difficulty arises from a subsequent irrational endeavour to interpret the latter expression as if it literally described what we conceive.

There is one position of *q* in which it does not cut *p*. 'In every other position *q* cuts *p* in some finite point.' What is a 'finite' point? Here, at all events, it is mere surplusage. We require nothing more than the plain, unambiguous statement: In every other position *q* cuts *p*.[1] On the other hand, 'if we move the point *A* to an infinite distance in *p*, then the line *q* which passes through *A* will be a line which does not cut *p* at any finite point.' Now this is not a literal, but a highly metaphorical expression of the process of thought which, literally, we may express somewhat as follows: If we suppose the point *A* to recede without limit in the direction *AB*, the line *q* which passes through *A* continues to approach indefinitely near to that position in which it does not cut *p*. Now if this does substantially and literally describe the process of thought—and I say 'if' because I cannot answer for any other mental process than my own—then it is clear that to be 'led to say that every

[1] The addition of 'in some finite point' is ingenious, because the alternative then gets itself expressed in so simple and innocent a manner by 'in some infinite point' or 'at some infinite distance'.

line through *S* which joins it to any point at an infinite distance in *p* is parallel to *p* ', and to find in this statement something contradictory of, or inconsistent with, Euclid's 12th axiom, is the result of comparing a literal with a metaphorical statement of the same process of thought, and, by taking the metaphor as literal, of concluding that there are two inconsistent processes of thought. What we are 'led to say' is just as metaphorical as the proposition which leads us to say it, and so soon as this is recognized the difficulty is seen to be gratuitous—there is no real contradiction to overcome.

Does this conclusion condemn the use of such an expression as 'the point at infinity in a line'? Not in the least. It condemns a semi-mystical, semi-sophistical validation of its employment. Its real justification lies in this: that it affords us a means of ignoring, without contradicting, the exception involved in Euclid's 12th axiom, and thus enables us to state the theorems of projective geometry with complete generality and yet without thereby contradicting any of the results arrived at by Euclid.

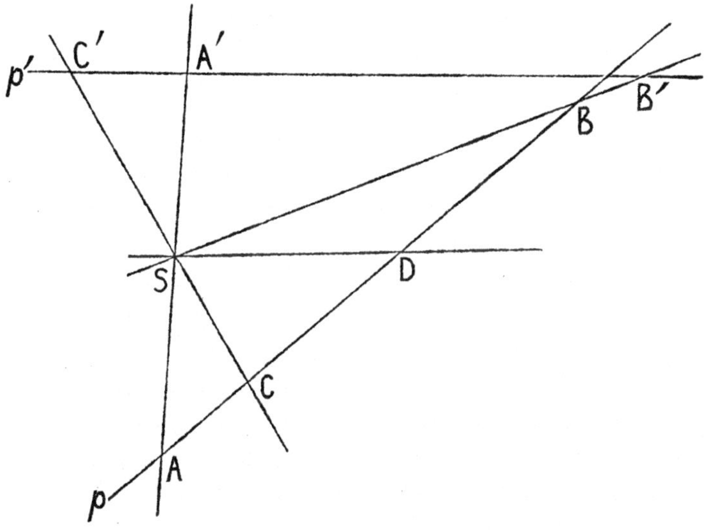

If from *S* we project the points of the line or base *p* upon the line or base *p*′, and speak of *A* and *A*′, *B* and *B*′, &c., as corresponding points, it is clear that there is one point in p—the point D where SD, parallel to p', cuts p—to which there is no corresponding point in p'.

This exception stands in the way of expressing with complete generality some of the theorems of projective geometry; for instance, the fundamental theorem that the anharmonic ration of any four points in a line is equal to the anharmonic ratio of their projections on any other line in the same plane. In the case just considered where D, one of the four points A, B, C, D in p, has no corresponding point in p', the theorem is, literally, inapplicable, and therefore cannot be said to be either true or false. But if we imagine a point D' in p' to recede indefinitely either in the direction $A'B'$ or $B'A'$, then the anharmonic ratio of the four points in p, the 'cross-ratio' $\frac{AC}{CB} : \frac{AD}{DB}$, is the limit to which the anharmonic ratio $\frac{A'C'}{C'B'} : \frac{A'D'}{D'B'}$ approximates indefinitely close as the point D' recedes indefinitely far. This is a *literal* statement of a geometrical property or relation. I apprehend that mathematicians, whether they clearly recognize the fact or not, express this same geometrical property in the *metaphorical* statement that the anharmonic ratio of the points A, B, C, D in p is equal to the anharmonic ratio of the points A', B', C', and *the point at infinity* in p'. This metaphor or, to be more precise, purely verbal paradox, would lead us to say, if we had not already said it, that parallel straight lines meet at infinity—a proposition which is no more and no less non-literal than the one which, in the supposed case, would lead us to say it.

We see in the most unmistakable way that we have here the first step in the paradoxical expression of real geometrical properties or relations after the received manner—even to the consecrated phraseology. Replace the phrase 'the point at infinity in p'' by 'the *imaginary point* of intersection of p' and SD', and the proposition that the anharmonic ratio of the points A, B, C, D is equal to the anharmonic ratio of the points A', B', C' and the *imaginary point* of intersection of p' and SD, is a paradoxical mode of expressing the real geometrical relation

$$\frac{AC}{CB} : \frac{AD}{DB} = \frac{A'C'}{C'B'} : 1.$$

In the doctrine of geometrical imaginaries the mystical bias to which the judgement becomes subject through the reaction of symbolic forms, whether these consist of words or of other symbols, is very much less marked in its effect than in the theory of algebraic imaginaries: the reaction is in fact largely held in check because in pure geometry the process of reasoning is conducted mainly by means of representative images (i. e. images typical of the geometrical concepts reasoned about); the accompanying symbolic imagery being in general non-essential, at times even obstructive, as Mr. Francis Galton assures us to be the case with him. Nevertheless, in so far as the symbol (of whatever kind) is effectively used as instrument in geometrical reasoning, to that extent may the mystical tendency reappear and assert itself. If now in retrospect we take a general view of this discussion of mathematical imaginaries, both algebraic and geometrical, recall its broad features, and in the light of it compare the algebraic with the geometrical case, we shall, I think, be impressed with the very close analogy which it discloses between some of the forms of expression in algebra and the metaphorical or paradoxical forms of speech in geometry; and with the similarity of the mode of reaction on the judgement of the reasoner, although the scope of it is much less in the one case than in the other. This analogy is in many points so marked that we might almost describe the algebraic forms in question as metaphorical or (in those which involve $\sqrt{-1}$) paradoxical expressions of real quantitative relation.

To be led geometrically to the notion of imaginary points is, then, to be led to invent this expression 'imaginary points' as a means of indicating, briefly though paradoxically, that certain relations, in virtue of which lines or figures form a system, are independent of the contingency of intersection of the lines or figures. But Cayley also spoke of our being led analytically to the same notion. There are two general methods of dealing with geometrical theorems and problems: the method of pure geometry and the method of co-ordinate or analytical geometry. In arriving geometrically at the notion of imaginary points we are in the domain of pure geometry; in arriving analytically at the same notion we are in the domain of analytical geometry. It is, then, tolerably plain that the literal meaning of being led analytically to this same notion of imaginary points, is the being led to turn the language of analysis or algebra to the same paradoxical use in analytical geometry as we have turned ordinary language in pure geometry.

We have seen how in algebra such combinations of symbols as $\sqrt{-1}$, $a+bi$, &c., are wrenched from their contexts and considered as algebraical expressions, although when thus isolated they have no definite meaning, or are paradoxical expressions which do not indicate relation of, or operation with, abstract quantity or number. These combinations of symbols are said to receive an interpretation in geometry. Argand, as we saw in the last chapter, devised a method of using them as literally symbolic of certain geometrical relations. In the case we are now considering they are used as paradoxically symbolic of certain geometrical relations, viz. those relations which, in pure geometry, are paradoxically expressed in ordinary language. If this is not what we are analytically led to, then what we are analytically led to are mystical propositions.

The peculiarity of Cayley's explanation, of his position, is that he seemed unable or unwilling to cut himself completely loose from mystical implications. This is clearly indicated by the question: What is the meaning of an imaginary point? with which he brings to an end his explanation. Let us now reconsider this explanation:

'In the Cartesian geometry a curve is determined by means of the equation existing between the co-ordinates (x, y) of any point thereof. In the case of a right line this equation is linear; in the case of a circle, or more generally of a conic, the equation is of the second order; and generally, when the equation is of the order n, the curve which it represents is said to be of the order n. In the case of two given curves, there are thus two equations satisfied by the co-ordinates (x, y) of the several points of intersection, and these give rise to an equation of a certain order for the co-ordinate x or y of a point of intersection. In the case of a straight line and a circle, this is a quadric equation; it has two roots, real or imaginary. There are therefore two points of intersection, viz. a straight line and a circle intersect *always* in two points, real or imaginary. It is in this way that we are led analytically to the notion of imaginary points in geometry.'

The simplest equation of a circle is obtained when we select, as axes of co-ordinates, two straight lines which intersect in the centre of the circle and are perpendicular to one another.

If we say that a is the length of the radius, the equation of the circle is $x^2+y^2 = a^2$. Any straight line in the plane of the circle, and referred to the same axes, will be 'represented' by the equation $y = mx+c$, where c is the length of the intercept on the axis of y, and m is the ratio of this length to that of the intercept on the axis of x; m is, in other words, the tangent of the angle which the straight line in question makes with the axis of x.

If the straight line and circle intersect, in the literal or ordinary sense of this word, say in the points A and B, then we can find the co-ordinates of these points by treating the equations of the circle and the straight line as simultaneous equations; that is, the points A and B being common to the two lines, we can substitute for the x (or y) of the one equation the equivalent of x (or y) furnished by the other equation. We thus get the quadric equation to which Cayley alludes. In this case, the circle being represented by $x^2+y^2 = a^2$, the abscissae of the two points of intersection are given by the equation $x^2+(mx+c)^2 = a^2$. The roots of this equation (the abscissae in question) are, for every case of intersection (in the ordinary sense of the term), real roots; moreover, intersection is the condition of simultaneity of the two equations. Cayley, however, says that these roots are either real or imaginary. To admit this carries with it that the term 'intersection' is from the outset of the explanation employed both in a literal and in a paradoxical sense. But if we thus read the passage it amounts to this: that we are led analytically to the notion of imaginary points by starting with that notion.

The notion seems to be rather that of denoting imaginary intersection by imaginary co-ordinates analytically expressed. If we pretend that the circle and straight line intersect always (which pretence we express by saying that they necessarily intersect either in real or in imaginary points), then it is a formally logical consequence of this pretence that the equations of the circle and straight line may always be treated as simultaneous. The roots of the equation $x^2+(mx+c)^2 = a^2$ are $\pm\frac{\sqrt{a^2(1+m^2)-c^2}}{1+m^2} - \frac{mc}{1+m^2}$. If the circle and straight line intersect in the literal sense the values of a, m, and c are so related that the quantity under the radical sign is positive and the roots are real. If, on the contrary, the circle and straight line have no points or point in common, that is, if they intersect in imaginary points, the quantity under the radical sign is negative and the roots are imaginary: these are, therefore, in virtue of the pretended premise, the abscissae of the imaginary points of intersection.

The process, thus far, may be said to exhibit the mathematician at play. To regard it as furnishing a geometrical interpretation of algebraic imaginaries in the usual sense of the word interpretation, or in any sense that would ultimately prompt the question, What is the meaning of an imaginary point? would be to take a *jeu d'esprit* seriously, or to be unconscious or oblivious of the fact that the process is a subtle game in which we play at interpretation, pretend to interpret. If we are conscious of this fact we shall not ask this question; nor shall we suppose that we have found an interpretation in pretending to find one. The game has 'serious scientific value' only if, when we say that the roots of this quadric equation symbolize either real or imaginary abscissae, we can in this paradoxical phraseology call attention to some geometrical property involved in the system of the circle and straight line which is independent of their having, in any point or points, co-ordinates in common, independent, that is, of the straight line being secant or non-secant of the circle. The process, then, expressed algebraically so as to involve forms of the type $P+Qi$ as co-ordinates of imaginary points of intersection, is a paradoxical mode of intimating that we consider the straight line and circle in some such relation or relations. If we look into modern textbooks of geometry we find that the straight line, whether it cuts the circle or not, is called a 'polar', or, to be more precise, the polar of a point with respect to the circle. This point is called the 'pole' of the straight line. Given the circle and the straight line, the pole is found by means of the same construction, whether the straight line intersects with the circle or not; it lies on the perpendicular from the centre of the circle to the straight line, either inside or outside the circumference, according as the straight line is non-secant or secant (if the line is a tangent the pole coincides with the point of contact), and the system has the permanent property that the square of the radius of the circle is equal to the product of the distances of the pole and polar from the centre of the circle.

A literal definition of polar and pole may run as follows: If in the straight line any two points be taken outside the circle, the two pairs of tangents drawn from them determine two chords which intersect either inside the circle or outside it in their prolongations. This point of intersection is called the pole of the straight line, and the straight line is called the polar of this point. But in paradoxical language the definition is much briefer. The chord determined by the tangents drawn from a point outside the circle is the polar, or lies in the polar, of the point. If we pretend that a circle and straight line intersect always, and express this by saying that the points of intersection are either real or imaginary, we can also pretend that tangents to the circle can be drawn from any point whether inside or outside it, and then define the polar of any point as the straight line which passes through the (real or imaginary) points of contact of the (real or imaginary) tangents drawn from the point. As a matter of fact we do find the polar of a point with respect to a circle defined in this paradoxical phraseology as well as in customary language.

When mathematicians say that algebraic imaginaries receive an interpretation in geometry, they mean apparently just what Cayley meant when he said that we are analytically led to the notion of imaginary *loci*: they do not, it seems, refer to Argand's 'representation of imaginary quantities in geometrical constructions' or to any developments of Argand's method. But, be this as it may, the contrast between the two methods of employing these algebraic forms in geometry calls attention to the very unusual and strained sense in which the term 'interpretation' itself is used in connexion with these forms and imaginary *loci*. For in a quite ordinary and readily understood sense Argand does give geometrical interpretations of these algebraic forms. In his method such combinations as $\sqrt{-1}$, $a+bi$, become definitely symbolic in the ordinary sense of the word. In analytical geometry, and in the same sense, they do not; that is, they are not, in the ordinary sense of the word, interpreted. An inquirer or learner, on being told that these forms, though uninterpretable as symbols of quantity, are interpretable in geometry, would on examination find this statement natural and intelligible with respect to Argand's invention, but forced or fanciful with regard to co-ordinate geometry—

supposing always that he did not succumb to mystical suggestions. Yet were he asked what more suitable, or less inappropriate, term could be suggested, he would find himself in a dilemma; for since the intention is to indicate some real geometrical property or properties by the employment of algebraic imaginaries, these properties have the formal right to be called interpretations of the algebraic imaginaries, however far-fetched (which does not mean useless) the term may seem. The connexion between symbol and meaning being recognized as essentially artificial, there is no ground for objecting to its being thus used. Paradox in the use of symbols becomes an artifice or convention in expression which, like all other conventions in the use of symbols, receives ultimate sanction only in virtue of its fulfilling some useful purpose.

The use to which Argand put the imaginaries of algebra in symbolizing geometrical conceptions is literal and direct; the use to which these algebraic forms are put in analytical geometry is paradoxical and indirect. In so far as the latter is wittingly, deliberately paradoxical (i.e. in so far as the mode of expression is intended and clearly recognized as non-literal and not even metaphorical), the conditions for a mystical judgement of the process are absent; and the indirect reference to real geometrical properties turns the process into something more than a mere play upon symbols or make-believe interpretation.

PART III

METAGEOMETRY

CHAPTER X

WHAT IS GEOMETRY?

Ambiguity of the expression 'Properties of Space'.—Geometry as the science of Configuration.—The alleged incertitude of the Axiom of Parallels in relation to the conceivability of different Kinds of Space.—Are the fundamental notions of Geometry particular ideas or general ideas?—Particular ideas and general ideas necessarily involve one another, and neither class can be more fundamental than the other.—Technical use of the terms 'Definition' and 'Indefinables' by mathematicians.

It is still customary, though perhaps less so than it was a few years ago, to define geometry as the science or the investigation of the properties of space. The definition is itself, however, of comparatively modern origin; it was not inherited from the geometers of antiquity; it has been remarked that the term space itself is not to be found either in Euclid or in Archimedes. If we look into textbooks or treatises of geometry we find that the talk is of points, lines, surfaces, solids; straight lines, circles, conic sections; planes, cylinders, cones, &c.: in general, of shapes or figures simple by comparison with those of the vast mass of common objects which surround us—so that if it were deemed desirable to give some brief definition of geometry, one would suppose that such a definition, at once brief and intelligible, would be that geometry is the science or investigation of shape or configuration in some of its simpler aspects. This is a sufficient indication, to begin with, of the subject of thought; for, in the end, definition of any subject of thought waits upon the development of the subject.

The science or investigation of the properties of space must always have appeared, to some thinkers at least, somewhat of an absurdity—even a contradiction, in so far as their conception of space, in contrast with that of body or matter, involved the negation of properties. How far this must in general be the case may be judged if we reflect that the ether of space was invented in order to escape the inconceivability of 'action at a distance' (an inconceivability which is quite genuine save when the term 'action' is emptied of its meaning, which is what

J. S. Mill did in his discussion of the question); and this inconceivability is at bottom none other than that involved in assigning properties to space. I do not know when this definition first came into vogue, but I should imagine we owe it principally to the peculiar development of geometrical speculation which began with Gauss, and which received from him the name 'non-Euclidean'. In the course of that development, geometry, as it was known to Euclid, became one of several possible geometrical 'systems', each of which was associated with, was involved in or involved, a particular conception of space. Such phrases as 'kinds of space', 'species of space,' 'varieties of space' (whether expressive of real conceptual development or of mystical illusions) once admitted, would naturally lead to the use of the term 'properties' in relation to space, for already in the use of such terms as 'kinds', 'species,' 'varieties,' space has been assimilated, however distantly, with body. If it is urged that these terms are thus employed metaphorically, the answer is inevitable; metaphor in the use of words expresses either the awareness of an analogy in thought, or the belief that we have, or ought to have, this awareness. In neither case can we escape the consequences, though we may sometimes succeed in shutting our eyes to them. They are very obvious in Helmholtz's essay 'On the Origin and Significance of Geometrical Axioms', especially where, in order to illustrate the differences between the Euclidean and each of the two principal non-Euclidean spaces, he admits the supposition of a moderately rigid body transferred from one kind of space to another (though the kinds cannot, of course, co-exist), and considers the changes of shape to which it would thus be subjected—an illustration which involves an obscure and yet persistent notion of space *acting* on body, which is a stultification of the notion of space itself.

Different self-consistent systems of geometry, if each system is to be exclusive of the others, appear to be possible only on the condition that we have several different notions of space,[1] each one exclusive of the others, and the *fons et origo* of one of these systems of geometry; for it is clear that not only the individual's first obscure awareness of space, but also the idea or notion arising from the perpetual process of comparison which goes

[1] Or, again, several different notions of metrical relation (Cayley's *Theory of Distance*).

on in this developing awareness, are pre-conditions of his geometrizing. From this point of view the emergence in the first half of the nineteenth century of different systems of geometry is not a little enigmatic, for, as a matter of fact, the origin of these non-Euclidean geometrical systems is attributed by the originators themselves (e. g. by Lobatschewsky in his *Studies on the Theory of Parallels*) not to any modification in the conception of space, but to the alleged incertitude involved in Euclid's so-called Axiom of Parallels. These systems then led, or were supposed to lead, to different notions of space. It may indeed be held that the uncertainty alleged to be involved in Euclid's axiom is itself the result of an indefiniteness in the notion of space, and that the development of self-consistent but mutually exclusive systems of geometry was a clearing-up of this indefiniteness, a vanishing of it in the establishment of definite but differing notions of space. To discuss that question is, in large measure, to discuss the philosophy of modern geometry.

A brief yet comprehensive view of geometry as it presents itself to a considerable number of contemporary mathematicians is given by L. Couturat in his *Principles of Mathematics*.[1] Mr. Couturat expressly disclaims any pretensions to originality, and this circumstance is, in his eyes, precisely what should recommend his work to the reader. He means by this that he gives expression therein not to an isolated set of opinions, but in the main to the opinions elaborated and largely held in common by a number of brilliant as well as patient modern investigators of mathematical principles. In original intention, as Mr. Couturat explains in his preface, the work was to be merely an account of the *magistral ouvrage* of Mr. Bertrand Russell,[2] but in the course of this exposition Mr. Couturat found himself gradually drawn to include in his account a brief analysis of the main part of the work of contemporary mathematicians on the same subject. We have thus in this useful work, and within the compass of some 200 pages, the pith and marrow of the new views on mathematical principles, the remaining pages of the volume being assigned to an interesting and damaging criticism of Kant's philosophy of mathematics.

'Geometry'—says Mr. Couturat—'is still commonly regarded

[1] *Les Principes des Mathématiques*, Paris, Félix Alcan, 1905.
[2] *The Principles of Mathematics*, vol. i, Cambridge University Press, 1903.

as being the science of space. According to sound method then, it would seem that we ought to begin with a definition of space. Now in the first place such a definition is very difficult and complicated, and in the next it is perfectly useless: the idea of and the very word space are not to be found either in Euclid or in Archimedes. The same may be said of the notions of the line and the surface, which Euclid himself attempts to define at the outset of his *Elements*. The definition of these general notions requires very great tact, and they become rigorous only by the aid of the integral calculus; and this is equivalent to saying that their place is neither in the elements nor in the principles of Geometry. It must not then be supposed that if these three notions cannot be defined at the outset of geometry this is because they are primary, fundamental, and simple; on the contrary, it is because they are very complex, and we can constitute geometry perfectly well without them, as will presently be seen. Geometry is founded, not upon the general and vague ideas of space, of surface, and of line, but upon the particular and precise ideas of the straight line, the plane, and especially the point; and it is among these that we find the primary and indefinable notions of this science. The point especially is the indefinable element of all systems of geometry. Points are the individual terms of all the relations the study of which constitutes the several geometries; and if space can be defined at the outset of geometry, it must be as the aggregate of points'[1] (l'ensemble des points).

It is clear from this passage that the terms 'definition' and 'indefinable' are not employed in accordance with general custom, but in a technical sense especially relevant to mathematics, approaching to, if not actually identical with, what is commonly understood by calculation or determination (cf. the allusion to the integral calculus in relation to the definitions of the line and the surface). If we were to agree that the idea of 'one' is simple and indefinable, we might then also agree that all finite integral numbers are definable by means of this indefinable, together with the idea (whether indefinable or not) of addition. But to force an interpretation upon a term which already has a recognized meaning, when there is another term whose recognized meaning is that forced upon the first, is a procedure which tends ultimately to confusion of ideas. And this does in some measure show itself, no matter which of the new expositions of elementary geometry we may turn to. The geometer finds himself obliged to give some definitions in the

[1] *Op. cit.*, pp. 126, 127.

ordinary sense; that is, to explain, by means of words whose meanings he assumes the learner to know, the technical sense in which the defined word or symbol is to be understood. But, starting also with certain notions which he calls indefinables, he is apt to suppose that it would be illogical to attempt to define them in the ordinary sense of definition; and, in so far as this is the case, there is confusion of thought, even if these notions should happen to be simple notions, and thus also insusceptible of any but purely nominal definition—that is, insusceptible of construction in the learner's mind by juxtaposition of simple notions; for definition serves not only this purpose, but also that of conveying information concerning customary—general or technical—use of terms; thus securing that we shall not be at cross-purposes in the discussion of any subject.

That we do not need to begin geometry by defining space is, of course, just a simple statement of fact; we understand Euclid without any need for a definition of space, or if we do not understand him it is not for the lack of it. Space defined as the aggregate of points makes one regret the more that the point is indefinable, for with at least one common notion of the point we seem in this way to express an incongruity of thought rather than a definition of space. The rather wild statement that not only the word space, but even the idea of space, is not to be found in Euclid, is a paradox into which Mr. Couturat appears to be hurried because he does not pause to reflect that conceptions do not alter or disappear merely because we extend, or alter, the meanings of names to suit our convenience or our theories. This, for instance, is what Mr. Russell does in his *Principles of Mathematics*, when he defines Geometry (p. 372) as the 'study of series of two or more dimensions', where 'dimensions, like order and continuity, are defined in purely abstract terms, without any reference to actual space' (p. 376). Mr. Russell is within his right, but Mr. Couturat is not within his when he affirms that *in Euclid* we do not find the idea of space.

Next, geometry is founded not on the vague and general ideas of space, of surface, and of line, but on the particular and precise ideas of the straight line, the plane, and the point. It must be remarked, however, that the very notions of the particular and the general involve one another, and that consequently nothing clear and definite is meant in speaking of an idea as particular save by reference to the idea or ideas which are general in relation to it. We do not escape this necessity in mathematics even though the particulars may form a continuous series or gradation of differences. For instance, equality is the idea of a particular relation of magnitude, but it is a condition of our having formed it that we shall also, *pari passu*, have developed that of inequality, which is a general idea subsumed, together with that of equality, under the still more general idea of magnitude. The same must be affirmed of the straight line and of the plane, neither of which particular ideas can possibly be conceived save by contra-distinction with the general ideas of the non-straight or curved line, of the non-plane or curved surface. It is thus incorrect to say, as Mr. Couturat says, that these *notions* or *ideas* are not necessary to constitute geometry, though it follows as a matter of course that if it is one of our purposes in geometry to define (in the arbitrary sense) these general notions in terms of the so-called indefinables, we must begin with the indefinables.

It may also be remarked that a general idea is not necessarily vague, unless by 'vague' general is meant, in which case 'vague' is surplusage; but a general idea may be more or less ample in content, and this will depend upon the extent to which the awareness of resemblance or analogy has been carried, either in actual experience or in constructive reminiscence of experience, or again, upon the extent to which assumption of resemblance or analogy has been carried in the pursuit of hypothetical explanations.

There is an insistence on the comparison between space, the surface, and the line on the one hand; the point, the plane, and the straight line on the other, which challenges attention and ought not to escape criticism. There is a clear analogy between the comparison of the straight line with the line, and the comparison of the plane with the surface; but this analogy completely breaks down when we come to the comparison of the point with space. The straight line is a kind of line, the plane is a kind of surface; the point is not a kind of space. The relation conceived between geometrical indefinables and geometrical definables seems thus to be lacking in definiteness, in coherence, and this betrays itself in the final statement that points are the individual terms of all relations, the study of which constitutes the different geometries. For if geometry is the study of the *relations* between points, then the fundamental and indefinable elements of geometry are to be found among these relations, and the point, instead of being 'especially' the indefinable element of geometry, is, on the contrary, formally excluded from among them.

I do not venture to offer any opinion on the merits of extending the meaning of the term geometry after the manner of the most recent school of philosophical mathematicians. The questions which I wish to discuss can be discussed without reference to this modern refinement, which must be looked upon as still on its trial; they are relevant to geometry as still commonly understood, in the narrower acceptation of the term. Nor does it much matter whether we set out with a formal definition of this geometry; though, as I have already remarked, if any definition is to be given, I should prefer 'the investigation of shape or configuration' as at least more descriptive than 'the science of space', for we might say of space—as the pious Erigena said of God—that it may not improperly be called nothing.

Ought we then, *en bonne logique*, as Mr. Couturat says, to begin by defining shape? Not so; it is neither good logic nor bad to leave it undefined; it is simply useless to attempt a definition which must trench upon the very investigation involved. The notion of shape is not only very general but very complex; a definition of it must give at least some analysis of that notion into its constituent notions; but this is just a part, and a most important part, of the study which is called geometry. It is true that it has been very much neglected, even by the quite modern school of mathematicians. It is, no doubt, convenient to take, e.g. the straight line as an indefinable, and to maintain a discreet silence upon the question whether this notion is fundamental or simple, not analysable into simpler notions; but the convenience appears to me to be obtained at the expense of an adequate philosophy of the subject.

CHAPTER XI

THE STRAIGHT LINE AND THE FLAT SURFACE

The terms 'straight' and 'plane' denote identity of linear and surface shape.—All other shape-names are class-names, i.e. names of shape-likenesses.—Elaboration of the notion of straightness and flatness; connexion with the notion of direction.—The *a priori* or transcendental in relation to Geometry.—The angle.—Linear shape and the notion of Length.—The notions of Direction and Length are fundamental in Geometry, even where the treatment of it is descriptive.

THE straight line and the plane surface are, as Mr. Couturat remarks, particular and precise ideas; that is, these names are names of a particular line, of a particular surface; but they are not names of one line and one only, of one surface and of one only. 'Straight' is the common name of many lines, 'plane' is the common name of many surfaces; they indicate respectively something in which the lines are alike, in which the surfaces are alike. That something is what we call shape of line, shape of surface.[1] Moreover, these names do not indicate merely likeness or resemblance, for if they merely did this, the notions of the straight line and of the plane would be general, not particular; they indicate identity of shape between the many lines, and again between the many surfaces. These identities constitute the particularities of the notions of the straight line and of the plane respectively. It is evident from this that the conception technically known as that of congruence is involved in that of a *particular* shape. To conceive distinct lines as of one and the same shape is to conceive the same linear shape in different places; and this is in essence the conception of congruence.

Although it cannot be admitted that geometry, in the ordinary sense, is possible without the *notions* of the line and surface—the particular notions of the straight line and plane being themselves impossible apart from these general notions—it is none

[1] It may not be out of place to remark that, in his criticism of Kant's philosophy of mathematics, Mr. Couturat himself insists on the notion of shape as fundamental: 'En réalité la ligne droite est une *figure*' (*Op. cit.*, p. 280; the italics are his).

the less a matter of fact that we begin geometry with these particular shapes of the line and surface. But as no one linear or surface shape is any more or less particular than any other, the mere fact that these are particular throws no light upon the question why we begin to geometrize with this particular shape of line and with this particular shape of surface rather than with any others, or why, of the innumerable linear and surface shapes, this particular shape of line and shape of surface have received proper names, all other geometrical names indicative of shape being without exception names of shape-likenesses, class-names. It is to be presumed that we find, in the linear shape which we call straight, and in the surface shape which we call plane, some very simple and characteristic property which stamps them respectively as standards of comparison with reference to all other linear and surface shapes.

When the straight line is said to be indefinable, it appears at least to be implied that this notion is not analysable, cannot be constructed by a synthesis of simpler notions, this impossibility clearly manifesting itself in the lack of real significance exhibited by attempted definitions; i.e. the condition of understanding the defining context is found to be that the formally defined notion must already have been developed. This is evidently what is meant, e.g. by the writer of the article 'Geometry, Part I,' in the *Encyclopaedia Britannica*,[1] when he remarks that 'Euclid's Definition 4, I—A straight line is that which lies evenly between its extreme points—must be meaningless to any one who has not the notion of straightness in his mind.' And the remark is no doubt true—for the mature individual. It might even be urged that it is rather the notion to which the name 'straight' is given which determines the precise sense in which the term 'evenly' is used in 'lying evenly between'. But it must not be forgotten that we elaborate many of our conceptions before we have made any effective acquisition of language, and that this prior elaboration is a necessary condition of the intelligent use of words and of the aid afforded by language in the growth of the conceptual system. Thus with regard to questions touching the genesis, order, and interdependence of those conceptions elaborated prior to the effective acquisition of language, the internal evidence afforded by language itself is superficial.

[1] Vol. x, p. 376.

Shape is an attribute of body, or the conception is, in the usual phraseology, an abstraction from body. But abstraction is, in fact, a process: the very definite process of withdrawing or abstracting attention, either in perception or in imagination, from some aspects of a complex presentation, and concentrating it on others, or on one. The aspects (which correspond in the main to the different avenues of sense) may be called abstractions, and the word then loses its misleading but common implications. Either perception or imagery of some kind is thus a condition of the process of abstraction in this definite sense of the word.

We perceive the shapes of body mainly through a combination of the sense of touch with that of the muscular adjustments which control the movements of the exploring organ. It is thus clear that a concurrent condition of the development of shape-perception is a development of the perception of situation or position in relation to the percipient. We may even go so far as to say that some rudimentary idea of direction and distance of objects from the percipient is already involved in the process of exploring the simplest shape, say that of an edge. This very brief description of the process of perceiving shape is applicable, by a significant metaphor, as well to the eye as it literally is to the hand, particularly with regard to the perception of edges or outlines. In the visual exploration of the shape of an edge or outline, the fovea, or spot of distinct vision in the retina is, metaphorically, kept 'in touch' with the successive points of the edge or outline.

But it is a loose, and even in some measure an inaccurate statement that we perceive the shape of an edge or outline by means of the muscular adjustments which determine the movements of the exploring organ, that we judge two edges to be of the same or of different shape according to the identity or difference of the muscular action in the two explorations. As a matter of fact we do judge two or more edges or outlines to be of the same shape notwithstanding very great differences in the muscular adjustments involved in the several explorations. On the table at which I am writing stands a box which has six distinct surfaces meeting two by two in twelve distinct edges, of which nine happen to be visible to me. I perceive that these nine edges are all of the same shape, but a very elementary knowledge of the anatomy of the visual apparatus as a whole is sufficient to show me that this judgement of identity in shape cannot proceed simply from the sense of the muscular adjustments in these nine distinct explorations; the muscular action in fact varies very largely from case to case, sometimes involving one pair of the three pairs of muscles which control the orbital movements, sometimes another pair, and again, and in general, various combinations in action of two pairs. If I perceive these nine edges to be identical in shape, this judgement must repose upon some identity of sign which is either involved in, or accompanies, these nine different muscular actions; and such a sign very obviously lies in the character of the movement impressed upon the eyeball in its socket by these muscular actions. If we consider this rotatory movement of the ball in its socket we see that there is a simple movement, which may readily become a standard of comparison, from which all other movements of the ball deviate more or less. This simple movement is rotation of the eyeball about a fixed or unchanging axis, all other or complex movements departing more or less from this simple type. It would, of course, be absurd to suppose that the geometrical conceptions involved in this description of the signs felt are present to the mind of the percipient. What we assume is that the signs *are* felt, and if we do not admit some such assumption we cannot account for an empirical derivation of our knowledge of the shapes of objects, we cannot even admit experience to be the source of this knowledge, and must conclude that this knowledge is *a priori*.

All edges or outlines, the exploration of which by the eye[1] involves this simple movement only, will eventually be judged to be of the same shape, no matter about what axis the rotation takes place, no matter what the amount of the rotation, no matter what the sense of it. This judgement reposes partly upon the ascertained fact that any edge, the exploration of which yields this sign, is found to yield it no matter where the percipient sees it from, partly from the conception of congruence or identity of shape, the same shape of line in different places or distinct lines of the same shape. But there is yet one more experience which is found to be peculiar to the exploration of this shape of line and to no other, viz. that if the end points of

[1] I confine myself to visual exploration because of the comparative simplicity of eye movements, and hence of the signs afforded by them.

the line are brought by the percipient into the same direction the sign vanishes completely—in other words, *all* the points of the line lie in the same direction; in Euclid's phraseology it 'lies evenly between its extreme points'.

I think we must admit that this shape is not fully characterized until this last result of its exploration has been achieved; in other words, that the conception of straightness and curvature of edge depends, for its complete elaboration, upon co-ordination with that of direction and change of direction. Uniformity of direction is what is implied in Euclid's 'lying evenly between'. But it is a question by no means easy to answer whether it is not in this very co-ordination that the conception of straightness and of uniformity of direction become definite.

Although it is mainly from the organic exploration of bodily edges that we elaborate the conception of linear shape, this is not the sole source of the conception. Evidently when we follow the movement of a small body the motion of the organ is identical with that which it would have if it explored some bodily edge. When we make abstraction of the size of the body, that is, consider it as a point, we are prompted by precisely the same motive which leads us to make abstraction of the body whose edge or edges we explore. Accordingly we get either, on the one hand, such a definition of the line as Euclid gives us, or, on the other, that the path of a point is a line; [1] and, corresponding to Euclid's 'statical' definition of the straight line, we should also have this definition in the form 'the path of a point which moves without change of direction'. From the orthodox point of view, however, the latter definition would probably at once be rejected on the ground that the straight line is already 'presupposed' in the conception of direction. It is a natural and obvious objection to raise that the very simplest form of the conception of direction is that of the mutual inclination of two intersecting straight lines. But it is equally evident from what has gone before that this view requires at least some modification. It is at all events clear that the conceptions of direction and of straightness originate in one and

[1] The confusion between the definition of a notion and an assumption or axiom which relates to the notion is carried to its extremity by Helmholtz, for whom such a statement as: the path of a point is a line, is the expression of an assumption or axiom. See his 'Origin and Significance of Geometrical Axioms,' in *Popular Lectures on Scientific Subjects*, vol. ii, p. 31.

the same matrix, in the same set of concrete experiences, and the utmost that can be conceded to the orthodox view is that these conceptions develop *pari passu* and only attain complete definiteness in their final correlation. In other words, we do not fully conceive the straight line until we have correlated it with direction, and it may be that in this very correlation the conception of direction is at the same time rendered definite. From this point of view it remains an open question whether we should consider the straight line as definable or not in the ordinary sense of definition.

From the moment the fusion of the straight line and direction has been effected, the familiar notion of the *line* of vision is constructed, and thus those similarities of sign, given in the movements of the eye, which come to stand for straightness of outline, now become symbolic also of flatness of surface. That is to say, the shape of surface swept out by the line of vision in exploring a straight outline becomes necessarily associated with this shape of outline. Hence one of the definitions commonly given of the plane, viz. that we construct it (in imagination) by drawing straight lines from all points of a straight line to any point not in the straight line (*Ency. Brit.*, art. Geometry, vol. x, p. 377).

To return to the connexion between the conceptions of straightness and of direction, I think it will be seen that the belief that the conception of straightness is a pre-condition of the conceiving of direction [1] is incompatible with any derivation of the former from experience, and carries with it the conclusion that this conception is *a priori*. This standpoint, or mental attitude, seems to be closely connected with that to which attention was drawn in chapter vi, I mean Cayley's standpoint as disclosed by his brief criticism of J. S. Mill's views on Geometry. Cayley saw very well, what curiously enough seems to have escaped Mill's attention, that 'if there is no conception of straightness, then it is meaningless to deny the existence of a perfectly straight line'.[2] But, for Cayley, straightness or the perfectly straight line,

[1] I do not know of any case in which this belief or opinion is actually expressed, but it appears to be implied in the use to which the term 'direction' is put, when employed at all, in geometrical treatises. It is certainly not commonly defined, nor introduced until after the straight line has been either defined or presented as an indefinable.

[2] Mill not only denied the existence of these geometrical entities, but also

and the other 'purely imaginary objects' of Mill, are 'the only realities, the ὄντος ὄντα, in regard to which the corresponding physical objects are as the shadows in the cave; and it is only by means of them that we are able to deny the existence of a corresponding physical object'. This is equivalent to saying that the purely imaginary objects of geometry are not derivable from experience, but are *a priori* in origin;[1] Cayley's own view of this *a priority* consisting apparently of a particular adaptation of Plato's doctrine of Ideas and of Reminiscence.

Now that we should adopt, or be driven to take refuge in, some such theory as this, as a consequence of the conclusion that the straight line and all the other purely imaginary objects of geometry do not exist in nature, or even as a result of the simple fact that we can ask such a question as whether they do or do not exist in nature—this is not only quite natural, but even inevitable, provided the fact that we come to this conclusion, or can ask this question, is incompatible with an empirical origin of these geometrical conceptions, i.e. the purely imaginary objects.

But this, in my opinion, is very far from being the case. It is not necessary, in order to be able to deny, e.g. that straight edges exist in nature, or that motion in a straight path occurs in nature, to possess an *a priori* standard of straightness. There appears to me to be no difficulty in admitting that such a judgement is perfectly compatible with an empirical origin of the conception of straightness. At the very least, doubt as to whether perfectly straight edges, or motion in perfectly straight lines, exist in nature, is eventually bound to be suggested through the simple and notorious fact that every one of our organs of sense is susceptible of being trained, even in adult life, beyond

that we can conceive them. What he meant by this it is hard to say: possibly that the mental images to which we give the names 'straight lines', 'circles', &c., are only replicas of the shapes actually existent; hence that straight lines, circles, &c., as geometrically defined, exist neither in nature nor in the mind. But even so the question remains: What enables Mill, or any one else, to make such a statement?

[1] The purely imaginary objects of geometry are not, on this view, confined to the 'indefinables' of geometry, as the mention of the circle, &c., shows. They must thus consist of the whole of geometrical indefinable and definable configurations. Geometry, in short, must be regarded according to this view as being wholly *a priori*. This is also Mr. Russell's position. Cf. *Mind*, July, 1907.

the normal acuity of perception. In infancy and early childhood the individual's mental development is largely bound up in learning to use his organs of sense, in learning to perceive as well as to act. Similarities and dissimilarities which are not perceptible to him at one stage of development become perceptible at a later stage. Admitting some such process of exploration and perception of linear shape as we briefly suggested above, it then becomes perfectly intelligible, as we shall presently see, that a judgement such as Mill's and Cayley's, as to the non-existence of straight lines in nature, is compatible with an experiential origin of the conception of straightness.

It must be borne in mind that we carry with us no recollection of the long laborious process of learning to perceive. Thus it is not until we begin to philosophize that the question presents itself whether the entities of geometry exist in nature. The individual who is either too practical or too busy to philosophize accepts without further reflection, or has accepted at some time without reflection, that the edge of a mathematical ruler is perfectly straight. He is then invited to look at the edge of such a ruler through a powerful microscope, and he perceives it to be not perfectly straight. Is it necessary to assume that he has an *a priori* conception of straightness which enables him to conclude that even the edge of a mathematical ruler is not perfectly straight? Evidently not; he perceives the edge as not straight through the microscope by contrast with his perception of it as straight by the naked eye. The microscope does here for the mature individual, on a very large scale and in a brief moment, what, on a very much smaller scale and in a much longer period of time, was effected for him in his immaturity through the training of his visual apparatus. An untrained thinker, for the first time making this experience with the ruler and microscope, would not improbably assert that the edge of the ruler is quite straight, though it is 'rough'; and the conception of uniformity of direction which we all possess would at once enable us to understand what he meant by the assertion. The judgement would of course be a rash one, for we could feel no assurance whatever that the edge of the ruler might not be after all slightly curved, and that the microscope may be as ineffective to make us sensible of the curvature as the naked eye is to make us aware of the roughness.

From the moment we realize, be it even only in a perfunctory manner, how the conception of straightness can be elaborated out of the material of experience, we are no longer intellectually justified in assigning to this conception an *a priori* origin as explicative of our possessing it. To do so would be very much like persisting in the explanation of the existence of natural species by the hypothesis of special acts of creation, after Darwin and his followers have made it plain that these species may perfectly well have originated not supernaturally, but naturally. But whoever admits the straight line and the other 'purely imaginary objects' of geometry to originate in experience, rejects the theory that geometry is an *a priori* science; like all other 'pure' sciences, it is a process of reasoning about abstractions from experience, about abstract ways of regarding the concrete. Admit that in nature there exist no straight lines, planes, spheres, &c.; you admit also that in nature there exist no rigid levers, inextensible strings, frictionless pulleys, &c. These latter notions also originate in experience and by processes very similar to those which yield the geometrical entities. Elementary theoretical mechanics is a process of reasoning about these and similar abstractions, just as geometry is a process of reasoning about the others.

The consciousness of that break, in the continuity of linear shape, which arrests attention in the course of the exploration of bodily edge or outline, develops into the conception of the angle, rectilinear and curvilinear; and, in accordance with the view here advocated of the origin of the fundamental geometrical conceptions, those of the rectilinear angle and of difference of direction complete and establish one another so soon as the percipient has correlated the straight line with uniformity of direction. Hence also there is, properly speaking, neither rectilinear nor curvilinear angle: the 'break' is rectilinear, or curvilinear, or partly one and partly the other, while the angle is the difference of initial direction, or, if the lines are straight, the inclination of the one to the other. But, the straight line and uniformity of direction having been correlated and in a loose sense identified, we find the so-called curvilinear angle defined as formed by the meeting of the tangents to two curved lines at their point of intersection (cf. *The Century Dictionary*, under 'Angle').

Clifford, in his *Common Sense of the Exact Sciences*, tells us that 'shape is a matter of angles', and illustrates this proposition in various ways. The statement is manifestly inapplicable to linear and surface shape. But we might say, employing the term figure in a restricted sense, that figure is a matter of angles. In this sense the solid has no shape, other than that or those of its surfaces and edges, but it may have figure. The point, however, is of little importance; and it would certainly be inconvenient thus to restrict the meaning of figure. Euclid's definitions of the line, surface, and solid are definitions of dimension rather than of figure.

We may remark, however, with reference to Euclid's definition of the line, that it appears to involve a conflict of opinion between him and some modern geometers. The line, says Euclid, is length without breadth, and—so we must read it—without thickness. Whatever we may think of this as a definition of the line (it shows us, of course, that he is defining an 'abstraction'[1]), we can at least gather from it what was Euclid's conception of length. Length, for Euclid, is a feature or attribute of all lines. According to this view the notion of length is not what it is, e.g. for Mr. Russell, 'originally derived from the straight line, and extended to other curves by dividing them into infinitesimal straight lines.'[2] Euclid's definition is quite general and precedes that of the straight line. What Mr. Russell describes as extending the notion of length from the straight line to other curves, is the notion of comparing a straight line and a curve in length by the intermediation of a broken line, the comparison becoming more and more precise as the number of breaks increases. But it is obviously a pre-condition of this notion, that the straight line, the broken line, and the curve are conceived under this common attribute, viz. length; we could not otherwise think of thus comparing them. The notion is, at bottom, the breaking of the straight line so that it becomes more and more nearly congruent with the curve. 'In the limit' of this process the length of the broken line is the length of the curve. No doubt,

[1] I do not by any means assert, however, that for Euclid the geometrical entities which he defined were nothing but abstractions. Euclid was by philosophical profession a Platonist, and, for aught I know, the geometrical entities thus defined may have been for him, as they were for Cayley, among the 'only realities'.

[2] *The Foundations of Geometry*, p. 17.

but 'in the limit' of this process the broken line is just the straight line bent into congruence with the curve. It is obviously because we conceive all lines, whether perceived or imagined, under this common attribute which we call length, that we have been led to the invention of the tape as well as of the wand. That the straight line, being the simplest shape of line and unique among linear shapes in the possession of a proper name, should have become in geometry the standard to which all other linear shapes are referred for length was inevitable, and is perfectly compatible with the fact that the notion of length is necessarily involved in that of the line, while the derivation of the notion of length from that of the straight line in particular, and its extension to other lines, is incompatible with that fact.

Length being a common, but simple and unanalysable attribute of lines, it follows that linear shape, with the exception of the straight shape, which involves uniformity of direction, must be analysable into relations of length and change of direction. In other words, non-straightness or curvature of line is, in analysis, a function of length and change of direction. But change of direction in the motion of a point describing a line is itself a complex notion, unless we confine the moving point to a plane. In this simple case differences of linear shape, whether we compare distinct lines, or different parts of the same line, are definable as differences in the rate of change of direction, length being the independent variable. The analysis of plane linear shape thus closely resembles that of velocity. Just as we have velocity, uniform, variable, and 'at an instant', so we have linear shape or curvature, uniform, variable, and 'at a point'. The usual geometrical way of analysing curvature of line springs from the customary definition of the circle, and is no doubt simpler in expression, but it hides from us rather too easily that the conception of change of direction is involved in that of curvature of line, and where the curvature is variable does not any more than the former avoid the notion of limit, which is contained in that of the circle of curvature 'at a point'.

The conclusion which, in terminating this chapter, I wish particularly to emphasize, is that the conceptions of direction and of length are fundamental in geometry, even where, as in projective geometry, it is not our intention to investigate either general or particular relations of direction or of length.

CHAPTER XII

DEFINITIONS AND AXIOMS IN GEOMETRY

Self-evidence and the object of Demonstration.—Discordant views regarding the distinction between Definitions and Axioms.—The Assumptions alleged to be hidden in Euclid's Definitions.—The Sense in which Geometrical Entities may be said to exist.—Irrelevance of Assumption to this sense of existence.—Mr. Poincaré and Professor Klein on the nature of Geometrical Axioms.—Similarity of Klein's views and those of Cayley.—Euclid's geometrical Axioms and Postulates.—Alleged distinction between the infinitude and the unboundedness of Space.—Some of the propositions usually classed as geometrical axioms are definitions of geometrical abstractions, in regard to which neither assumption nor convention has any relevance.

'EVERY conclusion rests upon premisses. These premisses are either self-evident and require no demonstration, or else they can be established only by derivation from other propositions. Since we cannot thus proceed *ad infinitum*, every deductive science, and geometry in particular, must be founded upon a certain number of indemonstrable axioms.'[1]

Robert Simson in the Notes to his edition of Euclid's *Elements*, tells us that

'Proclus, in his commentary, relates that the Epicureans derided Prop. 20' (any two sides of a triangle are greater than the third side) 'as being manifest even to asses, and needing no demonstration; and his answer is, that though the truth of it be manifest to our senses, yet it is science which must give the reason why two sides of a triangle are greater than the third; but the right answer to this objection against this and the 21st, and some other plain propositions, is, that the number of axioms ought not to be increased without necessity, as it must be if these propositions be not demonstrated.'

I do not know that Simson's answer is any better than that of Proclus. Some modern geometers reject, or believe that they reject, all appeal to intuition in geometry. 'L'intuition ne doit avoir aucune part réelle dans les raisonnements géométriques . . . ceux-ci, pour être rigoureux, doivent être purement logiques,'

[1] Poincaré, *La Science et l'Hypothèse*, p. 49.

says Mr. Couturat in his criticism of Kant;[1] and Mr. Russell, in his *Principles*, speaks of 'the mass of unanalysed prejudice which Kantians call intuition' (p. 260). But also, one of the requisites for the axioms, according to Mr. Whitehead, is that 'the total number of axioms should be few' (*The Axioms of Projective Geometry*, Cambridge Tracts in Mathematics and Mathematical Physics). I do not propose to say anything more here about the rôle of intuition in the axioms of geometry; but Simson's answer, it is evident, is a rule laid down without reason assigned. It may be a very good rule, but if we ask, What is its rational sanction, why should one self-evident proposition be admitted as an axiom and another not? the answer is not by any means obvious. The Epicureans say, in effect: The proof of a geometrical proposition consists in showing that it can ultimately be deduced from one or more elementary propositions which we accept as true. But it would be absurd to argue that we accept these elementary propositions as true because they cannot be proved; we accept them as true because they are self-evident. The rational object of demonstration, therefore, cannot be to establish the truth of self-evident propositions, self-evidence being the ultimate criterion of the truth of a proposition; the rational object of demonstration is to establish the truth or untruth of propositions which are not self-evidently true or false. Unless a valid reason is assigned for the rule, latter-day Epicureans will go on deriding these propositions as being manifest even to asses—which is their polite way of deriding as asses those who are at the trouble of demonstrating these propositions.

The modern geometer, however, has different sets of axioms from which flow different systems of geometry, and he is content to affirm, as Mr. Russell explains (cf. *Principles*, p. 373), merely that the set of propositions P is implied in the set of axioms A, the set P' in the set A', and so on, without asserting that this or the other set of axioms is the true set. The relevance of self-evidence to the axioms thus becomes somewhat vague, for although the inception of this process of inventing different systems of geometry lay in the negation of Euclid's axiom of parallels, on the ground that this axiom is not self-evident, yet the development of the process is represented as due to the desire of geometers to trace the consequences of rejecting other of the traditional

[1] *Les Principes des Mathématiques*, p. 288.

axioms (Ibid., p. 373). But what are we to conclude, on this particular point, from the result of these investigations? The three main systems of geometry—Euclid's, Lobatschewsky's, and Riemann's—are each self-consistent, but are, or are said to be, each inconsistent with the others. The axioms then, as Helmholtz observes,[1] cannot be 'necessities of thought'. Evidently, to accept this conclusion is to admit that the self-evident is not necessarily true. Thus, even from the Epicurean standpoint, that in geometry we are concerned with the truth or falsehood of propositions, it is a sound rule that we must admit in our geometrical premisses the least possible number of axioms. But it may be observed that we are in no need of the modern development of geometry in order to rationalize this rule, which appears to have been at least tacitly admitted ever since geometry became a methodical science. The fact is that we are all liable on occasion to admit, as self-evident, propositions which we subsequently recognize not to be such; so that finally we require a non-subjective or 'ejective' test, viz. the common consent which guarantees us against the stray illusions of the individual; the self-evident propositions are the residuum. If these propositions refer—to use Hume's terminology—not to matters of fact but to relations of ideas, their truth resides in the permanence of this 'ejective' self-evidence, and obviously cannot lie in anything else.

Geometrical axioms, no matter in what treatise or textbook we may look for them, are always propositions concerning some of the more elementary or fundamental of geometrical conceptions. If any definitions of these conceptions are given, these statements commonly appear to be of a different kind from the axioms. But it seems to be possible for the geometer to introduce, unguardedly and in the guise of definitions, statements which are in reality axiomatic propositions. Thus it is alleged that a number of axioms are hidden among Euclid's first definitions. It is a matter of some importance, then, to know precisely what, in the modern geometer's mind, is the distinction between an axiom and a definition. But this is also what it is extremely difficult to find out, because the modern geometer, if we generalize

[1] 'Origin and Significance of Geometrical Axioms.' *Popular Lectures on Scientific Subjects*, vol. ii, p. 54. Translated by Prof. E. Atkinson. London, 1881.

him, does not appear to know, while if we take him in his individual capacity we find, on the one hand, that for one distinguished living mathematician the axioms of geometry are nothing but definitions in disguise (Poincaré, *La Science et l'Hypothèse*, p. 66), while on the other, for the author of the article Geometry, Part I, in the *Encyclopaedia Britannica,* many of Euclid's definitions are axioms in disguise or 'assumptions'. There is evidently some grave difficulty about the question. Other modern mathematicians, no doubt because they recognize the difficulty, evade it. Thus Mr. Whitehead (*The Axioms of Projective Geometry*): 'Here "Definition" will always be used in the sense of "Nominal Definition", that is, as the assignment of a short name to a lengthy complex of ideas.' This, I believe, is more commonly called naming, but perhaps the name being short and the ideas long makes some difference.

Let us consider first the views of the author of the article 'Geometry' in the *Encyclopaedia Britannica* (vol. x, pp. 376, 377). 'The axioms are obtained from inspection of space and of solids in space, hence from experience. The same source gives us the notions of the geometrical entities to which the axioms relate, viz. solids, surfaces, lines or curves, and points.' Abstraction from experience thus appears to yield at once the geometrical entities and the propositions concerning them which are called axioms. This seems incompatible with the axioms being judgements, or inferences concerning the geometrical entities. We cannot, at any rate, get a synthetic judgement out of mere abstraction. Geometrical axioms are assumptions, as we see by the following excerpt:

'The assumptions actually made by Euclid may be stated as follows:

'1. Straight lines exist which have the property that any one of them may be produced both ways without limit, that through any two points in space such a line may be drawn, and that any two of them coincide throughout their indefinite extensions as soon as two points in the one coincide with two points in the other. (This gives the contents of Def. 4, part of Def. 35, the first two Postulates, and Axiom 10.)

'2. Plane surfaces or planes exist having the property laid down in Def. 7, that every straight line joining any two points in such a surface lies altogether in it.

'3. Right angles, as defined in Def. 10, are possible, and all right angles are equal; that is to say, wherever in space we take a plane, and wherever in that plane we construct a right angle, all angles thus constructed will be equal, so that any one of them may be made to coincide with any other. (Axiom 11.)

'4. The 12th Axiom of Euclid. This we shall not state now, but only introduce it when we cannot proceed any further without it.

'5. Figures may be freely moved in space without change of shape or size. This is assumed by Euclid, but not stated as an axiom.

'6. In any plane a circle may be described, having any point in that plane as a centre, and its distance from any other point in that plane as radius. (Postulate 3.)'

From this enumeration of the axioms contained in Euclid's preliminary statements, we ought to be able to gather precisely what, in the author's mind, constitutes the difference between the definition of a geometrical entity and an axiomatic proposition concerning it. Euclid, it is clear, defines these geometrical entities by means of what the author of the article calls their properties, and it is also fairly obvious that what the latter considers to be hidden in these definitions are the several assumptions that this, that, and the other geometrical entity exist. But it is in the first place an assumption itself that Euclid assumes this. It is, indeed, quite possible that Euclid, as a Platonist, looked upon e.g. the straight line and the plane as among the only 'real existences'. But Euclid the geometer is very careful not to mix up Plato's metaphysical speculations with the scientific exposition of geometry, as we clearly see from his procedure; he entirely ignores the question of the origin of these geometrical notions; he simply describes them. As Mill points out and as Cayley admits, these entities do not exist in nature. Nor is it necessary for the purpose of geometrical reasoning to assume that straight edges exist or that motion in straight lines ever actually takes place. The conclusions of geometry are significant if these fundamental notions are clear, and they are valid if the derivation is logical.

But if we do not and need not assume, in order to reason about these entities, that they exist in nature, do we assume that they exist in the mind? The question seems to have no sense. The existence of the entities consists in our conceiving them, and I fail to see what the relevance of assumption can be here. If it is said that what is assumed is that the entities have the described properties, this is merely a confusion of thought arising from the metaphorical use of the term 'properties'; the so-called properties by which the entities are defined are no other than the entities, i.e. the abstractions or conceptions themselves.

Mr. Poincaré reminds us that the word 'existence' has not, with reference to a mathematical entity, the sense which it has with reference to a material object. A mathematical entity exists provided no contradiction is involved in the definition of it, either in itself or with propositions previously admitted (*La Science et l'Hypothèse*, p. 59). Does the writer of the article in the *Encyclopaedia Britannica* mean, then, that in the definitions which he says hide axioms, it is assumed in each case that no contradiction is involved? We have to look behind the definitions to the things defined, which here are abstractions. What is the relevance of contradiction to the definition of an abstraction, if we exclude mere ignorance or inadvertence in the customary use of words? And hence what is the relevance of assuming that the definition does not involve contradiction?

Let us consider now in what way Mr. Poincaré reaches the conclusion that the axioms are definitions in disguise. He discusses the question in the third chapter of the book already mentioned, and begins by admitting the usual distinction between axioms which are essentially geometrical and those others of which Euclid's Axiom 1 (Things which are equal to the same thing, &c.) is the type. These Mr. Poincaré does not consider to be geometrical propositions but analytical propositions. 'I regard them as analytical judgements *a priori* and shall not concern myself with them any further.'

Although Mr. Poincaré follows here an almost universal tendency among mathematicians, it seems to me rash to hurry propositions of this kind out of sight without pausing to consider whether they may not throw at least a side-light on the nature of the specifically geometrical axioms. For we do not bestow names at random, but are always guided in the bestowal by the sense of some analogy, be it important or trivial, essential or superficial, and there must therefore be some reason, good or bad, for the assigning of this name 'axiom' to these propositions, as well as to the specifically geometrical ones. Mr. Poincaré does not attempt to explain or justify his view that these propositions are analytical judgements *a priori*, and I believe this view to be untenable. Euclid's Axiom 1, whether we call it a geometrical proposition or not, is a proposition about geometrical entities, simple or complex, fundamental or derived. 'Things' is a general term which covers lines, surfaces, solids, figures, angles, lengths, areas, contents, &c. Mr. Poincaré expresses the axiom in the form 'two quantities which are equal to the same third quantity are equal to one another', and, in relation to geometry, 'quantities' has the meaning which Euclid attributes to 'things'. Now this does not express an analytical judgement, and that this is so can very easily be seen if the process of thought which prompts it is expressed in a manner which does not hide the nature of that process, thus: Given (1) that A (any thing in Euclid's sense) is equal to B, and (2) that B is equal to C; then (3) A is equal to C. Here manifestly we have two data (1 and 2) and a conclusion (3) which follows from the data. That is to say, the conclusion results from a bringing together or synthesis of the data; we are enabled by means of this synthesis to arrive at something not given, and this is the essential trait of a synthetic judgement. The conclusion is at once immediate and apodeictic. If we are given the further datum that C is equal to D, the further conclusion that A is equal to D is mediate and apodeictic: it is mediated by the antecedent conclusion that C is equal to A. No one would dream of calling the first proposition a definition, and here at all events the term axiom has a clear and definite signification, viz. a synthetic proposition which embodies an immediate and necessary judgement or conclusion.

Returning from this digression to the main theme, viz. the case for geometrical axioms as definitions in disguise, Mr. Poincaré's argument is briefly as follows:

The three axioms which most geometrical treatises state explicitly are: (1) through two points one straight line only can be drawn, (2) the straight line is the shortest path from one point to another, (3) through a given point one straight line only can be drawn parallel to a given straight line. Numberless attempts were made to prove the third of these, known as Euclid's postulate, until Lobatschewsky and Bolyai showed that, starting from Euclid's other axioms but rejecting this one, a self-consistent geometry could be deduced inconsistent with Euclid's; hence that his postulate cannot be derived from the other axioms admitted by him. Subsequently it was shown by Riemann that a third system of geometry, equally self-consistent with the two others, but inconsistent with either of them, can be constructed if we reject, not only Euclid's postulate, but also the axiom that no more than one straight line can lie between two points.

The explicit axioms are, however, not the only ones, for after abandoning them successively there still remain a few propositions common to the systems of Euclid, Lobatschewsky, and Riemann. These propositions must be founded upon some premisses which geometers admit but do not state; and Mr. Poincaré finds these in the definitions of the straight line, of the plane, and of the equality of two figures; the last of which involves the notion of superposition and hence the assumption that figures can be freely moved without change of shape or size. We thus somewhat unexpectedly find ourselves back at the standpoint of the writer of the article on geometry in the *Encyclopaedia Britannica*, viz. that some of the definitions are axioms in disguise, or hide axioms. And by a geometrical axiom we see here that Mr. Poincaré means an assumption, something not absolutely self-evident as the case is with an analytical judgement *a priori* (cf. Ibid., p. 60).

Finally the nature of geometrical axioms is discussed, the question being asked, first, whether they are, as Kant affirmed, synthetic judgements *a priori*. They are not. If they were, they would impose themselves upon us with so overwhelming an authority that we could not conceive their contradictories, nor on these found different systems of geometry. Let us take, for instance, a real synthetic judgement *a priori*:

'If a theorem is true for the number one, and if we have proved it true for $n+1$ provided it is true for n, then it will be true for all positive integral numbers. Try to found, on the denial of this proposition, a false arithmetic analogous to non-Euclidean geometry. It cannot be done; one is even tempted at first sight to regard these judgements as analytical.'[1]

[1] The temptation is even greater to regard them as confusions of thought. Take, for instance, the second premiss by itself: If we have proved the theorem true for the positive integral number which immediately follows *any* positive integral number, provided it is true for *any* positive integral number.—How can this be a subject of proof? The entire proposition amounts to this:

If (1) a theorem is true for 1, and
 (2) can be proved true for 1+1, 2+1, 3+1, 4+1, . . . , provided
 (3) it is true for 1, 2, 3, 4, . . . , then
 (4) it is true for 1, 2, 3, 4, ...

Nor are the geometrical axioms empirical truths. We do not conduct experiments on ideal straight lines and circumferences, but only on material objects, that is, on their properties. Moreover, were geometry an experimental science it would not be an exact science, it would be subject to continual revision ; indeed, we ought to condemn it as erroneous, knowing as we do that no absolutely invariable solid exists.

'Geometrical axioms are therefore neither synthetic judgements *a priori* nor empirical facts. They are *conventions* ; our choice from all the conventions possible is *guided* by empirical facts, but it remains *free* and is limited only by the necessity of avoiding all contradiction. Thus the postulates remain rigorously true even though the empirical laws which have determined their adoption are approximative. In other words, the axioms of geometry (I do not speak of those of arithmetic) are only definitions in disguise.'

Thus, while for Professor Henrici (the author of the article Geometry in the *Encyclopaedia Britannica*) some of the definitions of geometry are axioms in disguise, for Mr. Poincaré, on the contrary, the axioms of geometry are definitions in disguise. We could hardly have more striking evidence of the looseness of thought—with its almost inevitable accompaniment, ambiguity in expression—which characterizes the modern discussion of geometrical principles.

What is at least quite clear is that the propositions which are commonly called geometrical axioms are of a wholly different type from that of the proposition known as Euclid's Axiom I, which embodies a conclusion, at once immediate and necessary, arising from the synthesis of two data or premisses. So far as the discussion has at present gone, these propositions embody abstractions from experience. This does not seem to be denied, not at least by Professor Henrici ; indeed, he affirms it. Nevertheless, in common with most mathematicians, he regards these propositions as embodying assumptions. Thus, in some unexplained manner, assumption becomes involved in the process of geometrical abstraction, and the consequence is that we are eventually landed in a dilemma. Instead of understanding that while Euclid's planimetry is an investigation of geometrical relations upon a surface of a particular shape, what is called Lobatschewsky's planimetry is a similar investigation with respect to any one of a class of surface shapes not contemplated by Euclid; and what is called Riemann's planimetry is a similar investigation in relation to any one of a class of surface shapes contemplated neither by Euclid nor by Lobatschewsky—instead of this, we are told that there are different possible geometries, corresponding to different possible spaces, according to the assumptions we make as to the properties of the plane, and that it is experience only, if ever, which can decide which shape is the real shape of the plane. Obviously, from this point of view, the plane is a surface which exists objectively but is not known to us with complete precision—or, at least, this should be so if the thinker is consistent in his process of thought. Mr. Poincaré at all events escapes from this dilemma. For him the question whether Euclidean geometry is true—which can only mean objectively true, since its self-consistency is not called in question —is meaningless. No one geometry can be truer than another; it can only be more *convenient*—a view of the matter which quite naturally follows from, or is consistent with, the opinion that geometrical axioms are conventions. At the same time it will be clear that both this view of geometry and this opinion as to the nature of the axioms are possible only for one who believes that the so-called non-Euclidean systems of geometry really possess geometrical significance, that we are really capable of conceiving different kinds of space.

Before returning to the subject of the 'assumptions made by Euclid', it will not be out of place here to give a very brief account of the views on the nature of geometrical axioms held by a distinguished authority, perhaps the greatest living authority, on the subject of non-Euclidean geometry, Professor Felix Klein.

According to Klein, the axioms of geometry give expression to a claim or demand of the intellect to transcend the limitations or the imprecision of our intuitions, a claim to absolute exactness. A figure in a geometrical demonstration should be looked upon as a means of making visible the sequence of its parts, the relations of position of points and lines, while leaving us aware that what we thus perceive inexactly or approximately is to be conceived as exact. This point of view implies considerable latitude in the choice of axioms. Every system of axioms should be admitted which, within the limits of exactness of intuition, is in agreement with intuition. It is to the inexactness of our spatial intuition that we must attribute the possibility of founding different systems of measurement, such as are embodied in hyperbolic, parabolic, or elliptic geometry. Intuition does not enable us to decide with certainty that any one of these three theories is false.

Turning now to the question of the origin of the axioms. If in these we recognize conceptual claims which transcend natural intuition, then the axioms cannot derive from experience. The results of actual measurement, however careful, are always subject to the degree of incertitude arising from the relative inexactness of intuition. But that we are able to consider these results in relation to absolutely precise axioms does not proceed from experience, but from a necessity of our own nature. Such a definition of the axioms implies that the foundations of mathematics are transcendental, reach beyond the imprecision of our intuitions. The import of the development of such a transcendental science in relation to its applications may perhaps be expressed thus: that in their application, the results of pure mathematics approximate to validity, the more exactly valid, in the application, are the premisses from which we set out on the path of mathematical deduction.[1]

In one respect—the question of the origin of the axioms—there is close similarity between Klein's views and Cayley's. What the former calls axioms are included among Mill's 'purely imaginary objects', which, according to Cayley, are the 'only realities', the ὄντως ὄντα, in regard to which the corresponding physical objects are as 'the shadows in the cave'. Cayley and Klein are thus at one in the supposition that geometrical axioms cannot derive from experience, and whether we look for their origin in the realms of the *a priori*, or of the 'only realities', or of the 'transcendental', seems to be more a matter of phraseology and the picturesque than anything else. But that the purely imaginary objects of geometry cannot originate in experience is what I am unable to admit. This question has already been discussed, briefly but, as it seems to me, sufficiently, in connexion with Cayley's views. Nor is it necessary to discuss Klein's definition of the axioms, since it is evidently suggested by the belief that they cannot be accounted for by experience.

[1] See *Nicht-Euklidische Geometrie*, part i, pp. 355 et seq.

We will return now to the consideration of the assumptions which, according to Professor Henrici, Euclid actually makes.

Euclid assumes that 'Right angles, as defined in Def. 10, are possible, and all right angles are equal; . . .' (Axiom 11).

This means that Euclid assumes that two straight lines can have the angular relation defined, viz. that one of them can stand on the other so as to make the adjacent angles equal. But since Euclid proves, in Props. 11 and 12, Book I, that this relation is possible, I do not see how he can be held to have assumed the possibility.[1] That the proof rests upon 'assumptions' does not warrant our counting this as a fresh one. The fact is the writer seems to be somewhat inconsistent in his treatment of Euclid. For instance, the list of assumptions said to be made by Euclid does not include parallels as defined in Def. 35. Why not? Evidently because in Prop. 27 Euclid 'proves the existence of parallel lines' (p. 378), i.e. proves this relation between straight lines to be possible. But if this is a good reason for not including Def. 35 among the assumptions, why not apply it in the case of Def. 10? The conception of the 'plane rectilineal' angle is general; it is the conception of angular magnitude or difference of direction between two straight lines in general. In Def. 10 (the right angle) Euclid defines a 'particular and precise' idea, i.e. that of a particular angular magnitude, just as in Def. 4 he defines a particular linear shape. Thus when in Axiom 11 he

[1] I put the case from what I conceive to be Professor Henrici's own standpoint, not from mine. I do not regard the notion of perpendicularity as something which we must either assume or demonstrate. I do not perceive the relevance either of assumption or of demonstration to the notion. The notion of a straight line which rotates in a plane about a point fixed in another straight line in this plane, involves the notion of perpendicularity, or equality of adjacent angles, as a particular case in the continuous series of cases; just as it also involves, as another particular case of the same series, that of coincidence, or disappearance of the adjacent angles (which we somewhat paradoxically express in saying that these become respectively equal to 0° and 180°). But although the notion of perpendicularity is necessarily involved as a particular case of the general conception of (linear) angular magnitude, it is quite a different matter to demonstrate that this geometrical relation (i. e. perpendicularity) is necessarily connected with other geometrical relations; and this is what, from a purely geometrical standpoint, Euclid does in the 11th and 12th propositions. It is not always easy, indeed, to analyse Euclid's procedure as purely geometrical, because he draws no systematic distinction between purely geometrical questions and questions which involve, in a greater or less degree, the notion of application or mensuration.

'assumes' that all right angles are equal, he 'assumes' that all angles of a particular magnitude are equal. The axiom is mere surplusage.

Let me return now to the consideration of Axiom 10—two straight lines cannot enclose a space, or one straight line only can lie between two points—which seems to me to be merely an alternative to Def. 4. This proposition

'has been regarded by some writers as either a mere definition of straight lines, or as contained by direct implication in the definition; but incorrectly. If it is held to be a definition, nothing is too complex to be so called, and the very meaning of a definition as a principle of science is abandoned; while, if it is said to be a logical implication of the definition, the whole science of geometry may as well be pronounced a congeries of analytic propositions. When straight line is strictly defined, the assertion is clearly seen to be synthetic.'[1]

This is not a little dogmatic. To say that if this proposition is held to be a definition, nothing is too complex to be so called, is mere rhetorical exaggeration; beating the big drum is not argument, but only drowns it. What proposition could well be less complex than the one in question? And the addendum, that the very meaning of a definition as a principle of science is abandoned, is but a begging of the question at issue, since it is precisely the difference between definition and axiom which is in dispute.

Lobatschewsky, in his *Geometrical Researches on the Theory of Parallels* (Berlin, 1840), gives a number of simple propositions (all in accord with Euclid) as premisses of his geometrical development. The first two of these propositions is as follows:

'A straight line is, in every position, superposable upon itself. I mean by this, that if about two points of a straight line we rotate the surface which contains it, this line does not change its place.'

Lobatschewsky merely gives this as a proposition. Is it a definition of the straight line, or is it an axiom? The two statements contained in it are evidently intended to be supplementary to one another, so that we may get at the exact meaning of the author. Evidently, also, a line which, in every position, is

[1] Art. 'Axiom', *Ency. Brit.*, vol. iii, p. 160.

superposable upon itself, is merely another way of saying, lines which, in every position, are superposable upon one another. The second statement is meant to give the precise sense of the 'in every position' of the first. It seems thus impossible to take this proposition as anything but a mere verbal variant of Euclid's Axiom 10. Yet the very next proposition which Lobatschewsky lays down is: Two straight lines cannot intersect twice. We cannot suppose that he intended to give the proposition twice over in different words. I conclude that he intended the first as a definition of the straight line, the second as an axiom about straight lines. Yet what difference, other than mere verbal difference, is there between them? Euclid's Def. 4 is equivalent to either of them; and so is also his Axiom 10.

I pass over Axiom 12, which will be considered further on under the theory of parallels; I also leave aside the assumption which Euclid is said to make *sub silentio*, that figures may be freely moved in space without change of shape or size. The discussion is rather long; it will form the subject of a separate chapter. There remain the three Postulates, about which I have as yet said nothing.

Some writers have seen in Euclid's Postulates merely statements which limit the use of instruments in geometrical construction to the ruler and compasses. But this is a side issue which has nothing to do with the foundations of geometrical reasoning. Whether the figures in which we embody or represent our geometrical conceptions are actually drawn or only imagined, and, if drawn, are free-hand or engineered no matter by what auxiliary instruments, has clearly no bearing on the kind of questions we have been discussing.

It is known that Euclid himself grouped the 10th, 11th, and 12th 'axioms', together with the three postulates, into one class under the name of αἰτήματα, and that the arrangement of the preliminary propositions actually found in the modern editions of his *Elements* was the work of his successors; the ground of the alteration being 'the distinction between postulates and axioms which has become the familiar one, that they are indemonstrable principles of construction and demonstration respectively'.[1] This distinction seems to be no longer accepted, at least by the quite modern school of geometers; and it is on

[1] Art. 'Axiom', *Ency. Brit.*, vol. iii, p. 159.

the whole strange that it should ever have been generally admitted; for, accidents of phraseology apart, it seems to be a distinction without a difference. Thus 'Axiom' 10, if expressed as it very commonly is expressed, viz. one straight line only can be drawn between two points, might just as well be called a principle of construction as a principle of demonstration. Suppose now that we say: A line which lies evenly between any two of its points will lie evenly between any two other points, this would be to state the so-called axiom of congruence for straight lines only, but is it to say anything in essence different from what Euclid says in the first postulate, that a straight line may be drawn from any one point to any other point? Personally I find no essential difference between the two; but this is no doubt due to the unusual interpretation I attach to the so-called axiom of congruence or assumption that figures may be freely moved, &c., which has yet to be discussed.

In the second postulate Euclid is said to assume that straight lines may be indefinitely produced; and, indeed, in this as in the other postulates, Euclid asks that something may be granted. But the particular form in which these propositions are cast may very possibly be due to the lack of a clear and consistent distinction between geometry and mensuration. It is at all events easy to see that if, in Euclid's day, that distinction was not observed, and it were simply asserted, e.g. that a circle can be described from any centre and at any distance from that centre, it would have been too easy for the critic to object that this is not possible. Euclid asks that this may be granted as a premiss: whoever will not grant it must go elsewhere for his geometry. But if we admit that in geometry we are concerned with what Hume called 'relations of ideas', while in mensuration our object is the application of these relations in 'matter of fact', which is the distinction in question, a distinction now very generally admitted by mathematicians, then to say that we assume the straight line to be indefinitely producible will be found to mean in substance this: that we assume the straight line not to be a circle. It is more than likely the reader may think that to call this an assumption is absurd or is an abuse of language, and that mathematicians cannot possibly intend anything of the kind. It may be they do not; but let us consider what we can make of the following passage, which is taken from Felix Klein's

Lectures on Non-Euclidean Geometry. Speaking of Riemann's conception of a constant 'measure of curvature' of space, he says:

'Riemann was here confronted with the question: what is the value of this constant measure of curvature? If it is equal to 0, we have the premisses of the ordinary Euclidean geometry. If it is negative we get Hyperbolic geometry, that is, the geometry developed by Gauss, Lobatschewsky, and Bolyai. But how stands the case if this value is positive? This possibility had so far been overlooked, or rather it had been put on one side because space had always and very naturally been taken as infinitely extended. But now Riemann observes that this third case can also quite well be admitted. For if it should indeed be that space is finite, that is, returns into itself, and that the length of the straight lines in which it returns into itself is merely very great, yet we can in no way become aware of the first of these properties, and we are therefore not warranted in neglecting this possibility. Thus while the earlier investigators naturally regarded space as infinite, Riemann says: space is indeed necessarily unbounded, but because space is unbounded it does not follow that it is infinite.'[1]

The Euclidean assumption is, then, that a straight line does not return into itself, and since it is a condition of metrical geometry that the unit of measurement is conceived as invariable both in shape and size, the Euclidean assumption is in essence that the straight line is not a circle. The conception of a line which returns into itself is simple; the phrase, space is finite, or returns into itself, is also simple; but does it give expression to any conception? Mr. Poincaré, like other eminent geometers, conceives two-dimensional beings living on a sphere. 'Their space will be unbounded, since on a sphere one can always go forward without being stopped, and yet it will be finite; you can find no end to it but can go round it'. Subsequently he tells us that 'Riemann's space is finite although unbounded, in the sense assigned above to those two words'.[2] Only what *is* this sense when we pass from a figure (which is a boundary in space) to space? Beings as 'two-dimensional' and a surface as a 'space' are merely misleading metaphors unless they express real analogies in thought. *Space*

[1] *Nicht-Euklidische Geometrie*, i und ii, von F. Klein. Zweiter Abdruck. Göttingen, 1893.

[2] *La Science et l'Hypothèse*, pp. 53, 54.

as finite but unbounded! Well might the least Mephistophelian of us exclaim:

> Denn eben wo Begriffe fehlen,
> Da stellt ein Wort zur rechten Zeit sich ein.

The third postulate completes the definition given of the circle in Def. 15. That is, the name 'circle' is not (like the 'straight' line) the name of a particular linear shape, but of a particular class of linear shapes.

It will be noticed that the trend of the argument developed in this chapter is in the direction of a conclusion which, up to a certain point, appears to be in agreement with that of Mr. Poincaré, viz. that geometrical axioms are definitions in disguise. But what do these definitions define? Apparently, according to Mr. Poincaré, they define conventions. But an abstraction from experience is not a convention in any ordinary sense of that word. Is it after all possible that Mr. Poincaré dissociates himself verbally only from the more common opinion that geometrical definitions are disguised assumptions? We have, according to him, a choice of conventions. Other geometers say that we have a choice of assumptions. According to the view maintained in the present work, these so-called axioms, or those at least which have so far been discussed, are simply definitions of geometrical abstractions in relation to which neither assumption nor convention have any relevance whatever. But we have not yet considered either the axiom of parallels or the axiom of congruence; and before we do so there is something more to be said about the axioms of magnitude.

CHAPTER XIII

THE AXIOMS OF MAGNITUDE

Analysis of Euclid's Axioms of Magnitude, taking Axiom 1, Book I of the *Elements* as typical.—Axioms 2, 3, 6, and 7, Book I, 1 and 2, Book V, are of the same type as Axiom 1, Book I.—Characteristics of this type of proposition.—These seven Axioms, though themselves general propositions, are subsumable under one or other of two more general and mutually converse axiomatic propositions.—Note on the use of the *reductio ad absurdum* argument with reference to Axioms 4 and 5, Book I.

IN the modern translations of Euclid's *Elements*, of the twelve propositions which, in Book I, are called Axioms, there are nine which are statements about magnitude. Several of these are known not to be authentic, that is, not to be Euclid's, but to have been added by subsequent editors or teachers. The historical aspect of these propositions I am not concerned with. I propose to take them as they stand, e.g. in Simson's *Euclid*, and inquire which of them are axioms in the sense that the first of them is an axiom, or conclusion which follows both necessarily and immediately from a synthesis of two data.

It is clear that the second and third of these propositions are axioms in this sense. We have

Given (1) the equality of A and B, and (2) the equality of C and D; with the conclusion $A+C=B+D$, and also $A-C=B-D$. Or, as Euclid expresses it:

Axiom 2. If equals be added to equals the wholes are equal. And

Axiom 3. If equals be taken from equals, the remainders are equal.

We have, next, the two propositions which deal with inequality:

Axiom 4. If equals be added to unequals, the wholes are unequal. And

Axiom 5. If equals be taken from unequals, the remainders are unequal.

Already Proclus had disallowed these two as axioms, on the

ground that they are derivative propositions.[1] And if derivation by means of the *reductio ad absurdum* argument is held to be in this case genuine reasoning,[2] the rejection is of course justified, these two propositions being respectively derivable by that form of argument from Axioms 3 and 2.

We come next to the propositions about double and half:

Axiom 6. Things which are double of the same, are equal to one another. And

Axiom 7. Things which are halves of the same, are equal to one another.

In the article 'Geometry', Part I (*Ency. Brit.*, vol. x), Professor Henrici tells us (p. 377) that a few of these 'common notions'—as Euclid called them—may be said to be definitions, and he suggests that Axiom 7 may be taken as a definition of 'halves'. But if so we might also take Axiom 6 as a definition of 'double'; and this would appear to land us in an absurdity. For since the defining contexts are in the two cases identical, it ought to follow that 'double' and 'half' have the same meaning. Similarly, for 'half' we might substitute 'a third', 'a quarter', &c.; and 'treble', 'quadruple', &c., for 'double'—the defining context remains throughout unaltered. The fact is, of course, that Axiom 6 is a particular case of the general proposition that equimultiples of the same magnitude are equal to one another; and Axiom 7 of the general proposition that those magnitudes, of which the same magnitude is an equimultiple, are equal to one another. These two general propositions are given by Euclid as axioms at the beginning of the Fifth Book of the *Elements*.

Now since a multiple is a sum of equals, it is clear that the first of these two general propositions (Axiom 1, Book V) is closely connected with Axiom 2, Book I. What is the nature of this connexion? Let it be admitted, first, that Axiom 2, Book I, is interpretable as follows:

Given that $A=B$, $C=D$, $E=F$, . . .; then $A+C+E+ \ldots = B+D+F+ \ldots$ Admitting this interpretation, then it is clear that the so-called Axiom 1, Book V, is merely the particular case of Axiom 2, Book I, where $A=C=E= \ldots$ But if we do not admit this interpretation of Axiom 2, in other words, if we restrict its significance to two pairs of equals, then the more

[1] Art. 'Axiom', *Ency. Brit.*, vol. iii, p. 159.

[2] See, on this subject, the note at the end of this chapter.

general conclusion must be held to be mediated by a series of syllogisms after the manner of Mr. Poincaré's mathematical induction, or demonstration by recurrence.[1] Thus if we have A, C, E given as equal respectively to B, D, F, the necessity of $A+C+E=B+D+F$ would follow only from a prior conclusion $A+C=B+D$.

I am quite unable to accept the latter view. Consider the still more general proposition: Magnitudes which are identically related to the same magnitude are equal. It is at once seen that Euclid's Axiom 1, Book I, is a particular case of this most general proposition; but it is no less clear that, given *any* identity of quantitative relation of X and Y to Z, whether it be equality or any inequality, the conclusion follows with the same immediacy and necessity.[2] In other words, this general proposition is axiomatic, and it clearly cannot be made more general.

The converse of this proposition, viz. Equal magnitudes are identically related to the same magnitude, is no less axiomatic and no less general. Between them they comprise every less general axiomatic proposition which can be made about equality. Euclid's Axioms 1, 2, 3, 6, and 7, Book I, and Axioms 1 and 2, Book V, are respectively subsumable under one or the other of the two.

From these two axioms of magnitude:

1. Magnitudes which are identically related to the same magnitude are equal to one another,
2. Equal magnitudes are identically related to the same magnitude;

we derive respectively, by the *reductio ad absurdum* argument, the propositions:

3. Unequal magnitudes are not identically related to the same magnitude,
4. Magnitudes which are not identically related to the same magnitude are unequal.

Euclid's axioms about inequality, viz. 4 and 5, Book I, 3 and 4, Book V, are comprised in these two derivative propositions.

[1] *La Science et l'Hypothèse*, chap. i. Mr. Russell's criticism of Mr. Poincaré on this point seems to me to be well founded. See *Mind*, July, 1905.

[2] It is, in fact, the given *identity* of relation, and not the *particular* relation, which carries with it the necessity of the conclusion.

This analysis of Euclid's axioms of magnitude plainly discloses one foundation of the opinion that the so-called axioms 8 and 9. Book I of the *Elements*, are definitions ; 8 of ' identical ' equality or congruence, and 9 either of the relation of whole to part, or of ' greater than '—as you please. Neither of these statements are subsumable under, or derivable from, the above general axioms 1 and 2.

The proof of a proposition by *reductio ad absurdum* of its contradictory is a mode of argument which Euclid occasionally uses in the *Elements*. It has always been regarded, I believe, as a valid process, and I have nothing to urge against it when employed as Euclid employs it, or on occasions similar to those on which he employed it. But with reference to Proclus's disallowance of Axioms 4 and 5 (sum and difference of equals and unequals) on the ground of their being derivative propositions, this mode of argument seems to me to sink to the level of a purely verbal artifice to which corresponds nothing more than the mere simulation of a process of judgement.

It is of course perfectly obvious that there is no difference of kind between any one and any other of the four propositions, Axioms 2, 3, 4, and 5. If any one of them contains an immediate and necessary conclusion, so does every other of them. We might just as well start with 4 and 5 as axioms and from these derive 3 and 2 by the same kind of proof. The admission of this kind of proof as being, in this case, really significant thus puts us in an absurd position. In the first place because when we say that 2 and 3 are axioms, and 4 and 5 demonstrable propositions, or vice versa, we imply that the connexion between data and conclusion in the one pair is not the same as in the other, which is not really the case ; in the second place because of the manifest incongruity involved in the very endeavour to mediate a conclusion which is actually an immediate conclusion.

If we consider the way in which the argument actually proceeds, it is quite possible to admit that the process has no real significance, and yet that the argument is one which, even in the particular case, it may be necessary to use.

Thus, having admitted as axiomatic the proposition : if equals be taken from equals, the remainders are equal, we proceed to prove the proposition : if equals be added to unequals the wholes are unequal. In the familiar Euclidean manner, but using algebraic symbols for brevity's sake :

Let $A=B$, and $C\neq D$, then shall $A+C\neq B+D$. For if not, suppose $A+C=B+D$. From these equals take the equals A, B. Then by the axiomatic proposition, it follows that $C=D$, which *ex hypothesi* is not the case. Hence the supposition that $A+C=B+D$ is false.

No doubt, but we have taken it for granted that the first step in the argument corresponds to a real movement of thought. Nothing is easier than to write, or say, ' suppose that $A+C=B+D$,' but the question is whether, with the data in mind, we *can* make the supposition. If we can, then the alternative $A+C\neq B+D$ is, so far, not necessary ; but then neither can the conclusion affirmed in the admitted axiom be necessary, and we have been mistaken in admitting it as such. I can see no issue from this dilemma other than to admit that the so-called demonstration is merely a verbal quibble.

But it may be considered as a quibble designed to encounter and defeat another. There have always been obstinately disputatious people, and others seeking reputation for wisdom through the expression of pyrrhonic doubt. The Greek geometer of old no doubt had to deal with such persons, for his countrymen were fond of argument. He had to deal, say, with a stubborn, perhaps a disingenuous, critic. The latter has been got so far as to admit the first three axioms, let us suppose, by means of the usual argument that no further discussion or development is possible without this prior admission. The 4th and 5th axioms are then proposed to him. Possibly he refuses to be drawn into any further admissions. But he has already gone too far, and the geometer holds his man. For if, by chance, his refusal is genuine, then he *can* suppose the alternative conclusion, and the proof tendered will have to be admitted, *by him,* as a genuine proof. On the other hand, if his refusal is pretence, he cannot object to the supposition without unmasking the pretence, that is, without admitting what he had just denied, viz. the necessary truth of the proposition.

It would seem, then, that demonstration by the mode in question is in these cases an argument *ad hominem* rather *ad judiciam.*

CHAPTER XIV

THE AXIOM OF PARALLELS

Form in which Euclid expresses the proposition.—It clearly embodies a conclusion suggested by given relations between geometrical entities.—This conclusion, whatever may be thought of its necessity, is certainly not immediate.—The same must be said of Playfair's version.—Conflict of opinion as to the self-evidence of the proposition.—Cayley's views on the question.—The proposition considered as expressing a generalization from experience, and hence as not apodeictic.—The two propositions into which the Axiom of Parallels can be broken up.—The converse of the first is geometrically equivalent to the second.—These two mutually converse propositions exhibit a remarkable analogy in thought-process with the two mutually converse Axioms of Magnitude of chapter xiii.—Suggestion that they are real Axioms of Direction.

EUCLID's Axiom of Parallels runs as follows: If a straight line meet two straight lines, so as to make the two interior angles on the same side of it, taken together, less than two right angles, these straight lines being continually produced, shall at length meet upon that side on which are the angles which are less than two right angles.

The reason which made Euclid put the proposition in this complicated form is plain when we compare it with the 16th, 27th (or 28th), and 29th propositions of Book I. The 27th follows from the 16th, the postulate is the converse of the 16th and serves to prove the 29th. The four propositions may be briefly expressed as follows:

Prop. 16. Two straight lines which meet are not identically inclined to a transversal.

Prop. 27. Two straight lines which are identically inclined to a transversal are parallel.

Postulate. Two straight lines which are not identically inclined to a transversal must eventually meet.

Prop. 29. Parallel straight lines are identically inclined to a transversal.

The so-called axiom, or postulate, is clearly not the definition of a geometrical entity, since it embodies a conclusion which, whether it does or does not necessarily follow from, is suggested by, the given relations between geometrical entities. Nor, were the conclusion admitted to follow necessarily from the premisses, could the proposition be admitted as an axiom in the precise sense in which I have used this term, viz. that of a conclusion which follows not only necessarily but also immediately from a synthesis of two premisses. That the conclusion, even if necessary, is not immediate, is evident from the fact that the proposition can be split up into two, the conclusion of the first becoming a premiss of the second:

1. Two straight lines which are in the same plane and are not identically inclined to a transversal are mutually inclined.

2. Two straight lines which are in the same plane and are mutually inclined must eventually meet.

In other words, the conclusion, whether necessary or not, is derivative. But since it is not admitted to be necessary, or self-evident, either one or both of the subordinate propositions is (or are) not necessary, or self-evident.

Euclid's statement of the postulate, as already remarked, lacks that simplicity and immediate intelligibility which should, if possible, characterize a fundamental proposition. This objection, however, cannot be urged against Playfair's version of the axiom, which is the one now commonly given: Through a point not in a given straight line there cannot be drawn, in the same plane with it, more than one straight line which does not cut it. But in substance, though more simply stated, it is the same proposition as Euclid's, and gives rise to the same two questions: (1) Is it a matter of doubt whether through the point in question there can lie but one straight line which is identical in direction with the other? If this is doubtful the conclusion as to intersection is doubtful. But if, on the other hand, it is a necessity of thought that identity of direction with a straight line *AB* is possible for one straight line only through a point *C* not in *AB*, then we have question (2), Is it doubtful that a straight line through the point *C*, in the same plane but not identical in direction with *AB*, must eventually cut it? Or, finally, are both these propositions doubtful? The reader should put the question to himself and answer it without an eye to the consequences which the answer may involve.

Mr. Russell seems to go somewhat too far in asserting that

honest people find it hard to assent to the axiom of parallels as self-evident.[1] It is, on the contrary, notorious that many honest people find it extremely hard not to do so, especially when the proposition is put as Playfair puts it. But I suppose these honest people are negligible by reason of their stupidity. The truth of the matter seems rather to be this: that 99 individuals out of every 100 who come to the study of geometry admit the proposition, when simply expressed, as self-evident; but when they hear that the premiss can be rejected and that a self-consistent geometry can be developed on the assumption that more than one straight line through the point (in Playfair's version) is parallel to the given straight line, i.e. is a non-secant, they are thrown into that state of 'momentary amazement and irresolution and confusion' which Hume characterizes as the result of pure scepticism; and if they study the arguments by which Lobatschewsky established this geometrical system, they find, as Hume found when he studied Berkeley's arguments, 'that they admit of no answer, and produce no conviction.'

Moreover, this question of self-evidence cannot be dismissed by merely waving aside as negligible the honest people who do find the postulate self-evident. Cayley is not a negligible quantity, and in his opinion Euclid's postulate does not need demonstration. This is one of the many puzzles with which the subject abounds, for Cayley was himself an original investigator and innovator in non-Euclidean geometry. The explanation which he himself gives of his standpoint is indeed not quite so clear as one could wish.

'My own view,' he says, 'is that Euclid's twelfth axiom in Playfair's form of it does not need demonstration, but is part of our notion of space, of the physical space of our experience—the space, that is, which we become acquainted with by experience, but which is the representation lying at the foundation of all external experience. Riemann's view before referred to may, I think, be said to be that, having *in intellectu* a more general notion of space (in fact a notion of non-Euclidean space), we learn by experience that space (the physical space of experience) is, if not exactly, at least to the highest degree of approximation, Euclidean space.'

The use of such expressions as 'physical space', 'the space with which we become acquainted by experience', seems to

[1] *Principles of Mathematics*, vol. i, p. 373.

imply that there are other spaces with which experience does not make us acquainted, or non-physical spaces, which we can conceive; and this would appear not to differ essentially from Riemann's view. We can, according to Cayley, have different systems of geometry in accord with the several different definitions which can be given of 'distance'; he does not clearly state whether we can or cannot have these several systems according to the several conceptions which, to follow Riemann, we can form of space. If we are to take what he says about the axiom of parallels for what it appears to mean, we must, however, conclude that he does not agree with Riemann, disbelieves that any other than the ordinary conception of space is possible.

I think it may rightly be said that in general the plain man admits Playfair's axiom as self-evident, while in general the mathematician does not admit it as such. I do not at all suggest that in such a matter we should count heads rather than weigh them; but the case is plainly not one of those of which it may be said that there is on one side popular delusion and on the other expert knowledge. Helmholtz remarks that we are all prone to mistake general results of experience for necessities of thought.[1] In his view the so-called axioms of geometry are not necessities of thought but general results of experience, that is, I suppose, what are commonly called 'laws of nature' or generalizations from experience. In taking the axiom of parallels as self-evident, the plain man, then, according to this view, confounds a generalization from experience with a necessity of thought. But, in the case of this axiom, the conditions of experience are such that it is not easy to see how the proposition can even present itself as the expression of a general result of experience. The certainty which we attribute to generalizations from experience, such for instance as those embodied in Newton's laws of motion, arises from our finding no exception to them within the limits of experience. But in the case of the axiom of parallels the empirical evidence is not quite of the same nature. In general we find that within the limits of experience the axiom is neither confirmed nor refuted, so that it is by no means clear how the supposed misapprehension can really occur. The judgement is certainly not plainly and unequivocally an empirical generalization.

[1] *Popular Lectures on Scientific Subjects*, Series 2, p. 32.

On the other hand, there is a suggestion of the empirical in the apparent element of prediction involved in the Euclidean (which is the classically original) form of the axiom. Given certain conditions, then something will happen, viz. the lines in question must eventually meet. It is by no means impossible that this mode of phraseology counted for something in forming the original judgement that the proposition is an empirical one. If this was so, the result for the philosophy of geometry was a very unfortunate one. It is easy to get rid of the apparent element of prediction contained in the so-called axiom by breaking it up into its two subordinate propositions, and by substituting for the second one the converse of the first:

(1) Two straight lines not identically inclined to a transversal are mutually inclined.

(2) Two straight lines which are mutually inclined, and are continually produced, will eventually meet.

Instead of (2), which contains the apparent element of prediction, we substitute the converse of (1), i.e. Two straight lines which are mutually inclined are not identically inclined to a transversal. From these two propositions, (1) and the converse of (1), it can be demonstrated that the sum of the angles of any triangle is equal to two right angles, a proposition which is equivalent to the postulate. But it is more convenient and neither more nor less self-evident to take as axioms the two propositions:

a. Straight lines which are identically inclined to the same straight line are not mutually inclined.

b. Straight lines which are not mutually inclined are identically inclined to the same straight line.

The proposition about the sum of the angles is seen at once to follow from these two.

The proposition (1) above and its converse derive respectively, by the *reductio ad absurdum* argument, from the propositions *b* and *a*, just as we have seen to be the case with the axioms of magnitude. In fact these four propositions relating to direction are in exact analogy with the four propositions relating to magnitude. Take, for instance, the first axiom of magnitude and proposition *a*:

If *A* and *B* are identically related in length to *C*, *A* and *B* are equal, or are identical in length.

The necessity is simply a suggestion, contained in the data, which we are unable to resist. To have the length-relation of *A* to *C* given as identical with that of *B* to *C*, suggests that it is indifferent whether we consider *A* or *B* in this relation to *C*; and it is this suggestion which imposes itself upon us as a necessity of thought: the identity of *A* and *B* in length is a necessary conclusion. Now put proposition *a* in a similar form:

If *A* and *B* are identically related to *C* in direction (are identically inclined to *C*),[1] *A* and *B* are identical in direction (are not mutually inclined).

Given the direction-relation of *A* to *C* as identical with that of *B* to *C*, suggests that it is indifferent whether we consider *A* or *B* in this relation to *C*. Those who find that this suggestion is one which they are unable to resist will, in other words, find that the identity of *A* and *B* in direction is a necessity of thought.

If we were to take proposition *a* as a *definition* of mutual non-inclination of two coplanar straight lines, it would follow (from my point of view) that there are no axioms in geometry other than the axioms of magnitude. But although this idea of the proposition as a definition is at first sight plausible enough, it will not stand careful examination. Evidently if we admit proposition *a* to be simply a definition of mutual non-inclination, we must then also admit the proposition: Two straight lines which are not identically inclined to a transversal are mutually inclined, to be a definition of mutual inclination. 'Mutual inclination' would thus be merely another name for the notion of non-identical inclination to a transversal. But it cannot be denied that the notion of mutual inclination is involved in that of the angle. Thus if the notion of mutual inclination is that of non-identical inclination to a transversal, the notion of the angle involves that of non-identical inclination of the straight lines which contain it to a transversal; and this is simply not the case.

It will be understood that the very close analogy evinced between the two sets of propositions respecting magnitude and direction is not urged as a demonstration of the self-evidence of the latter set. If self-evidence is the foundation of ratiocinative demonstration, the ratiocinative demonstration of self-

[1] Including non-inclination as a special case.

evidence is an absurdity. It is rather for those who already feel the necessity of the geometrical propositions that the precision of the analogy is in this respect of importance, for the very process of analysis which discloses the analogy affords a guarantee that the necessity already admitted is not an illusion due to over-haste in judgement.

CHAPTER XV

THE AXIOM OF FREE MOBILITY OR CONGRUENCE

This proposition is not necessarily implied in Euclid's process of reasoning. —The assumption that bodies can be moved without change of shape or size is relevant to Mensuration.—The proposition that geometrical figures can be thus moved is, as an *assumption*, meaningless ; it is merely one of several ways of defining the notion of Congruence.—Analysis of Clifford's explanation of this so-called geometrical Axiom.—Mr. Bertrand Russell's explanation of its meaning.

SUPPOSE some one had suggested—possibly some one did suggest—to Euclid, that in his *Elements* he assumes, but omits to state, that figures may be freely moved without change of shape or size. Might he not quite legitimately have replied that in geometry we are not concerned with the physical properties or qualities of bodies, with the effects of action ; such matters belonging to the province of physics, in which the results of geometrical investigation find a place as do those also of arithmetical investigation ?

By a legitimate reply I mean one consistent with the process of reasoning in the *Elements*. It will at once be objected that Euclid does make this assumption in the proof of Prop. 4, Book I, which may be said to be the fundamental proposition of the *Elements*. This opinion is general, but in order to maintain it effectively it seems that we must hold the so-called Axiom 8 (magnitudes which coincide with one another, that is, which exactly fill the same space, are equal to one another) to be the expression of an assumption. I see in it nothing but a definition of 'identical' equality or congruence. Professor Henrici says it 'may be taken' as a definition of 'equal'; and the writer of the article 'Axiom' (*Ency. Brit.*) already quoted, unhesitatingly pronounces it to be a mere definition of equals. But if that is so, the assertion that Euclid tacitly makes the said assumption in the proof of Prop. 4 may very well be disputed. In this proposition the premisses, or what is given as data for reasoning, are that two sides and the included angle of one triangle are equal to two sides and the included angle of another triangle; that is, these entities are given as coinciding on superposition; the conclusion is, that the third sides and the other angles must also coincide on superposition. Thus if we regard Axiom 8 as defining equality, the opinion is untenable that the demonstration (which consists in the deduction of the conclusion from the premisses) involves the assumption in question. Euclid, in this so-called axiom, defines coincidence as the exact filling of the *same* space (clearly a useless definition from the point of view of mensuration, or applied geometry); had he said *equal* spaces the term equal would obviously not have been defined in the proposition. Whence, then, do we derive the notion of equal, and hence also of unequal, *spaces*? Evidently from that of figure; the notion of equal spaces is here that of the same figure in different places: make abstraction of the notion of configuration, of delimitation in space, and the notion of spatial magnitude disappears.

In applied geometry, in mensuration, every precaution which experience suggests is taken to ensure, so far as possible, invariability of the instruments of measure—evidently because the conception of invariable figure is involved in that of metrical comparison. Thus there is here a clear sense in which it may be said that the validity of the results of measurement depends upon the truth of the assumption that the instruments of measure and the things measured have remained invariable during the process. But if the notion of invariable figure is involved in the very conception of metrical comparison, then, in pure geometry, which excludes every consideration subsumable under the conception of cause and effect, the statement that we *assume* invariability of figure seems to be bereft of any definite meaning. At first sight the statement that this is one of the axioms of metrical geometry seems so obviously true that it hardly occurs to us to question it, to inquire what, precisely, is the meaning of it. Geometers do not, as a rule, perceive that it calls for an explanation. In the few cases where such an explanation has been attempted, a careful analysis of it brings to light inconsistencies and confusions of thought. Let us take, for instance, that which Clifford gives in his book, *The Common Sense of the Exact Sciences*.

Geometry, according to Clifford, is a physical science, which deals with the sizes, shapes, and distances of things, and which can be developed by making two very simple and obvious observations and using these over and over again. These observations are:

First, that a thing may be moved about from one place to another without altering its size or shape.

Second, that it is possible to have things of the same shape but of different sizes.[1]

What Clifford here calls observations are what geometers usually call assumptions; and, as will be seen, it is as assumptions that Clifford really discusses them. We are concerned here with the first only of these two observations, which Clifford examines under the heading 'Lengths can be moved without Change'.[2]

He begins by remarking that this observation is the condition of the measurement of distance:

'The measurement of distance is only possible when we have something, say a yard measure or a piece of tape, which we can carry about and which does not alter its length while it is carried about.'

In the course of some explanations and illustrations of the process of measurement, which we need not reproduce at length, there occur the following remarks:

'Two lengths or distances are said to be *equal* when the same part of the measure will fit both of them.'

And again:

'We may say generally that two lengths or distances of any kind are equal, when one of them being brought close up to the other, they can be made to fit without alteration.'

The first rather than the second of these two statements seems at first sight to be intended as a definition of equality; but it becomes subsequently clear that the first is regarded as an assumption. Speaking of the comparison of two lengths by means of the tape, Clifford says:

'We find that each of them is equal to the same length of tape; and we assume that two lengths which are equal to the same length are equal to each other.'

In other words, if I understand him, the length *A*, coinciding with the length *B*, and being then applied to the length *C* and found to coincide with *C*, we assume that the length *C*, if it were

[1] Fourth edition, 1898, p. 47. [2] Ibid., pp. 52–5.

applied to *A*, would coincide with *A*. The lengths *A* and *C* being in different places, and supposing it to be either inconvenient or impossible to bring them together, we can only compare them, as Clifford says, by means of something which we can carry about and which does not alter its length while it is carried about. What we assume, according to Clifford, is that in carrying *B* from *A* to *C*, *B* remains all the while equal to *A*.

But what do we mean by saying that *B* remains all the while equal to *A*? That no matter from what place in *B*'s path from *A* to *C*, *B* were brought back to *A*, *B* would coincide again with *A*; or what comes to the same thing, that at whatever point of *B*'s path *A* were brought into apposition with *B*, they would again coincide.[1] In other words, the assumption is that *A* and *B*, travelling about in apposition, would remain coincident. But this assumption once made (supposing for the moment that it is relevant to geometry as distinct from mensuration), it is not a fresh assumption that *A* is equal to *C*; this follows as a necessity of thought from the data that *B* is and remains equal to *A*, and is also equal to *C*. The question is whether this proposition, 'Lengths can be moved without change,' a proposition which, in relation to problems of mensuration, clearly involves an assumption, has any meaning, *as an assumption*, in geometry. Clifford proceeds as follows:

> 'These considerations lead us to a very singular conclusion. The reader will probably have observed that we have defined length or distance by means of a measure which can be carried about *without changing its length*. But how then is the property of the measure to be tested? We may carry about a yard measure in the form of a stick, to test our tape with; but all we can prove in that way is that the two things are always of the same length when they are in the same place; not that this length is unaltered.'

What was clear in the previous explanation because we supposed we knew what Clifford meant by 'length' now turns to obscurity because we are no longer sure what he meant by 'length'. This much, however, seems to follow from the passage just quoted: that

[1] This seems to be what Clifford himself means, for he tells us that to assume that two lengths which are equal to the same length are equal to one another, is equivalent to assuming 'that if our piece of tape be carried round any closed curve and brought back to its original position, it will not have altered in length.'

by 'the length' of the tape we are not to understand its relation to the stick; for if, *ex hypothesi*, we find this relation to remain unchanged, it would be meaningless to ask whether it is changed. Similarly, if we were to take any other portable object as a test for the tape and were to ascertain that the relation of length between them remained unaltered, it would again be meaningless to inquire whether this relation was changed. In general, then, we must conclude that for Clifford 'the length' of an object is not a relation between it and any other object; and, with this conclusion, the previous explanation of the assumption that lengths can be moved without change becomes, to say the least of it, ambiguous. But let us go on to the 'singular conclusion':

'The fact is that everything would go on quite as well if we supposed that things did change in length by mere travelling from place to place, provided that (1) different things changed equally, and (2) anything which was carried about and brought back to its original position filled the same space. All that is wanted is that two things which fit in one place should also fit in another place, although brought there by different paths; unless, of course, there are reasons to the contrary. A piece of tape and a stick which fit one another in London will also fit one another in New York, although the stick may go there across the Atlantic, and the tape via India and the Pacific. Of course the stick may expand from damp and the tape may shrink from dryness; such non-geometrical circumstances would have to be allowed for. But so far as the geometrical conditions are concerned—the mere carrying about and change of place—two things which fit in one place will fit in another.'

It is the less necessary to discuss the singularity of this conclusion because in the very next sentence Clifford shows, apparently without being himself aware of it, that the singularity of the conclusion consists in its being meaningless. For he goes on to ask: 'Is it possible, however, that lengths do really change by mere moving about, without our knowing it?' and he concludes: 'Whoever likes to meditate seriously upon this question will find that it is wholly devoid of meaning. But the time employed in arriving at that conclusion will not have been altogether thrown away.' At last, then, after some ineffective hammering, Clifford hits the nail on the head. But, having arrived at this well-meditated conclusion, it is rather surprising that he failed to see that it makes an end of the singular conclusion previously arrived at; for this conclusion is, in fact, an answer to the question which he declares to be wholly devoid of meaning. The answer, in effect, is that either of the two assumptions, (1) that things do not change their length by mere moving about, and (2) that things do, under the given conditions, change their length by mere moving about, is consistent with what we know about moving things. But if we can give a rational answer to a question, the question cannot be devoid of meaning; on the other hand, if the question is devoid of meaning and nevertheless we answer it, some illusion of judgement must be involved in the process. I need scarcely add that I agree with Clifford in his final conclusion, i.e. that the question proposed is devoid of meaning. In other words, from the moment we make abstraction of all 'non-geometrical circumstances', free mobility without change of shape or size ceases, *as an assumption*, to have any meaning. Or, again, we may put it that what, in mensuration, is the assumption that two lengths which fit in any one place, and at any one time, will also fit in any other place, and at any other time—this, in geometry, or when we make abstraction of all 'non-geometrical circumstances', is the definition of equality of the two lengths. In geometry, as distinct from mensuration, it seems to me that free mobility, congruence, the same figure in different places, all express that notion in the absence of which the ordinary conception of metrical relations becomes obscure, and the expression 'metrical relations' cannot be used without ambiguity, unless redefined as the name of some conception which is not the ordinary one.

Mr. Russell, in his article on Non-Euclidean Geometry in the *Encyclopaedia Britannica*, recognizes that much confusion of thought exists on the subject of this so-called axiom or assumption of congruence. He attributes the confusion in large measure to neglect of the distinction between geometry and mensuration, but nevertheless holds that in geometry congruence is assumed. The axiom of congruence 'is the axiom in virtue of which superposition may be used (as in Euclid's fourth proposition) to prove the equality of two figures'. But the true meaning of this axiom is 'the existence of equal spaces, the fact that two spatial quantities may be equal or unequal'.[1] We see here the extreme elasticity of the term 'axiom'. Since the particular quantitative relation 'equality' and the complementary general quantitative

[1] *Op. cit.*, vol. xxviii, p. 671.

relation 'inequality' together constitute the content of the fundamental notion of quantitative relation or 'quantities', we seem here, under the guise of an 'axiom' that two spatial quantities may be equal or unequal, to be defining what we are thinking about under these terms.

Mr. Russell appears to have been led to this rather empty 'axiom' from the view, previously expressed in the same article, that motion is irrelevant to the foundations of metrical geometry:

'It is important to notice that we cannot appeal to *motion* in anything concerning the foundations of metrical geometry. Motion, as used in geometry, is not the motion of a single point, but the motion of many points, or even of all points. Now, if we ask how a motion is distinguished from other transformations, the only reply is, that it leaves metrical properties unchanged. . . . Thus metrical properties are presupposed in the definition of motion, and must not be defined by its means.'[1]

Unchanged metrical properties are presupposed in the definition of motion, but unchanged metrical properties itself presupposes the conception of motion; thus the argument is inconclusive. What Mr. Russell seems to have in mind is this: that when we speak of motion of a geometrical figure (not of a body), we cannot escape the necessity of defining a figure, in geometry, as one whose metrical properties remain invariable, otherwise the meaning of *a* figure disappears. And with this view I quite agree, but it is no other than this: A figure, in geometry, is a number of points which do not move relatively to one another, and the motion of a geometrical figure is the motion of a number of points which do not move relatively to one another.

The 'axiom' of congruence is then discussed in relation to the conception of rigid body:

'The definition of a rigid body, like the definition of geometrical motion, presupposes what is really meant by the axiom of congruence. A rigid body is one which, at different times, occupies equal spaces. Thus equality of spaces is logically prior to rigidity of bodies. How we discover two actual spaces to be equal is no concern of the geometer; all that concerns him is the existence of equal spaces, the fact that two spatial quantities may be equal or unequal. And this is indeed the true meaning of the axiom.'[2]

[1] *Op. cit.*, vol. xxviii, p. 670. [2] Ibid., p. 671.

That (as Mr. Russell affirms) 'rigid bodies are unnecessary in geometry' is a proposition which is not likely to be disputed by any one who draws a clear distinction between geometry and mensuration. But if, when he speaks of rigidity of body, the geometer has in mind invariability of figure—and this, I believe, is usually the case—then we have to do merely with a question of phraseology. Substitute invariable figure for rigid body, and the argument is a variant of the one which went before as to the irrelevance of motion; but it is no more convincing. I do not see how we can admit that the notion of invariable figure is derived from that of equal spaces. If any logical priority is to be assigned, I should assign it the other way; because, as I have already urged, if we make abstraction of figure altogether, the notion of spatial quantities disappears altogether.

We come back, then, to the conclusion already more than once expressed; that figures which coincide on superposition, or the same figure in different places, or motion of an invariable figure, are equivalent ways of giving expression to the notion of congruence. The *axiom* of congruence, using the term axiom in the definite sense of a proposition expressing an immediate and necessary conclusion which follows from the synthesis of two data not themselves conclusions, is obviously contained in Euclid's Axiom I, where the term 'equal' includes the meaning of 'congruent'.

CHAPTER XVI

SYSTEMS OF PLANE GEOMETRY

Real and Nominal Contradiction.—Conditions of Real Contradiction.—Euclid's and Lobatschewsky's respective hypotheses concerning parallels nominally exclude one another.—If these hypotheses are also real contradictories, we must admit two mutually contradictory planimetries for the surface which Euclid calls plane.—If, on the other hand, the two so-called planimetries are relevant to two different surfaces, both are admissible at the same time, and the contradiction is merely nominal.—The same argument applies to Riemann's planimetry in relation to Euclid's and Lobatschewsky's.—These conclusions are unavoidable unless it is a fact that different kinds of space are conceivable.

It is generally believed that Napoleon died at St. Helena. The assertion, which many people would be ready to make, that Alexander the Great did not die at St. Helena, would not be considered to be in contradiction with the belief that Napoleon died there; the one belief does not exclude the other; we find no difficulty in yielding credence to both assertions. But if *A* asserts that Napoleon died at St. Helena, and *B* asserts that Napoleon did not die there, the two assertions are in contradiction. It may not be the case, however, that the *beliefs* to which *A* and *B* thus give expression are irreconcilable: we might on inquiry discover that while *A*'s assertion was intended to be relevant to the great soldier and statesman, *B*'s was intended to refer to Victor Hugo's *Napoléon le Petit*. Between the two beliefs there would be no contradiction; any one could hold them both; the contradiction would be merely nominal.

It is no different when the predication relates not to indubitable events, but to really uncertain events, or to events to come. If I assert that Tom and Jane will meet here at dusk, this is not contradicted in your assertion that Charles and Mary will not meet here at dusk; nor would there be aught but a merely verbal contradiction, did you assert that Tom and Jane will not meet as aforesaid, you having in mind some Tom and Jane other than those to whom my assertion is relevant.

The case, again, is no different when, instead of events, we are dealing with abstractions. Euclid affirms that two straight lines in the same plane with, but not identically inclined to, a common transversal, must meet if continually produced. Lobatschewsky, in common with many other geometers, considers this proposition to be uncertain, or not self-evident. There are, according to him, two alternative and mutually exclusive hypotheses: either, as Euclid affirms, every straight line through *A* which falls within the right angle *CAB* will meet *BD* (which is perpendicular to *AB*); or else only those straight lines through *A* which fall within the acute angle *EAB* (of uncertain magnitude) will meet *BD*; the straight line *AE* being, on this side of *AB*, the limit between those lines which cut *BD* and those which do not cut it.

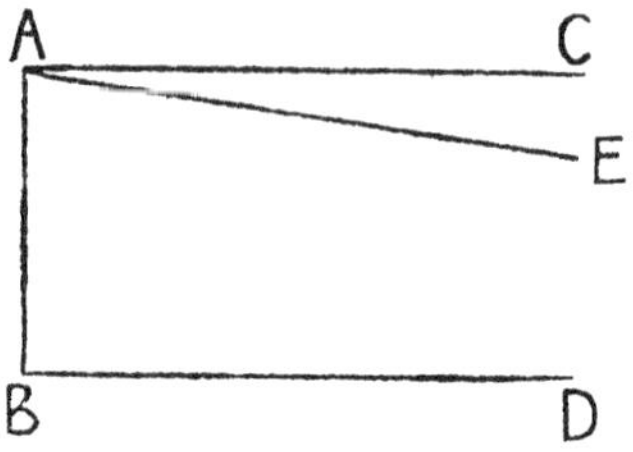

In admitting this second hypothesis Lobatschewsky at least verbally contradicts Euclid, or any one who affirms what Euclid affirms. But here, again, it is an underlying condition of the hypotheses being mutually exclusive, that is, of the contradiction being real, that the two assumptions shall be relevant to the same entities. If, for instance, we suppose Lobatschewsky to have meant, by a plane and a straight line, a surface and a line different from the surface and the line which Euclid had in mind —and by different I mean different in shape—then the two hypotheses are not mutually exclusive in the sense of standing in logical contradiction each to the other. To admit either of them need not necessarily involve the rejection of the other.

Professor H. J. Stephen Smith says, in his *Introduction to Clifford's Mathematical Papers*, that Lobatschewsky's assumption 'was in effect to adopt the hypothesis (though it does not appear to have occurred to Lobatschewsky in that light) that a plane has negative curvature'. But evidently no one can assume that a plane has negative curvature if by the term 'plane' he means a surface of zero curvature, which is what Euclid is understood to have meant, and is what people commonly do mean, by a plane. If it did not occur to Lobatschewsky that his hypothesis is equivalent to assuming that a plane has negative curvature, it is to be presumed that by a plane he did not mean such a surface, but meant the surface shape to which that name is commonly given ; and in that case his assumption is—as he seems to have intended it to be—a real alternative to Euclid's 12th Axiom, and stands in real contradiction to it. But if, on the other hand, Lobatschewsky did consider his hypothesis as equivalent to assuming that a plane has negative curvature, it is to be presumed that when he uses that term in his own development of geometry—Imaginary Geometry, as he calls it—he meant by it a surface of negative curvature ; and in that case his hypothesis is not—what he seems to have intended it to be—a real alternative to Euclid's, and does not stand in real contradiction to it. No one will find any difficulty in admitting that in a surface of zero curvature there is, or there may be, through a given point, but one parallel to a given straight line, while at the same time admitting that in a surface of constant negative curvature there are, or there may be, through a given point, an infinity of parallels to a given geodesic.

There is, I believe, not a particle of evidence in any of Lobatschewsky's published works which goes to show that by the terms ' plane ' and ' straight line ' he meant anything other than what we suppose Euclid to have meant by them. Indeed, if we cannot be sure of what Lobatschewsky conceived under these terms from the definitions he gives of them and the contexts in which he employs them, neither can we have any assurance, on the same subject, from the like evidence in Euclid's case. And in connexion with this point it may be well to call attention to a remark which Lobatschewsky himself makes in the preface to his small work on the *Theory of Parallels* (Berlin, 1840) : that in opposition to Legendre's opinion, the theory of parallels is wholly independent of such imperfections of principle as that of the definition of the straight line—blemishes, if blemishes they are, which he considers to be irrelevant to the theory in question. Short of explicitly affirming that imperfections in the definition of the straight line are of no moment because this conception, no matter how expressed, is perfectly definite, we could hardly have anything more explicit of Lobatschewsky's attitude than is this remark of his.

For many years after Lobatschewsky first made his researches, mathematicians in general regarded his non-Euclidean geometry as a logical freak, clearly because they supposed this geometry to be relevant to the Euclidean plane and straight line. The planimetry which he deduces from his rejection of Euclid's 12th Axiom contains many theorems which, as Mr. Poincaré remarks,[1] are strange and at first disconcerting. Why do they produce this effect upon us? Simply because we understand them to be predicated of the surface and the line which Euclid calls the plane surface and the straight line. Let it be understood that these theorems relate to the pseudo-spherical surface,[2] or surface of constant negative curvature, and we find nothing either strange of disconcerting in such a theorem as, e.g. that two 'straight' lines may be so situated in relation to one another that, being produced on the same side of a common transversal, they first approach one another and then recede from one another;[3] or, again, that two straight lines which are perpendicular to a common transversal continually diverge from one another on both sides of the transversal. These and many other theorems involved in Lobatschewsky's planimetry strike us as strange, and even as impossible, because we understand them to be affirmed of the particular shape of surface commonly termed 'plane', and of the particular shape of line commonly called 'straight'.

Bodies exist, and bodies have surfaces and, in general, edges and corners, in the ordinary senses of these terms imposed upon us by the practical affairs of life. But the surfaces, lines, and angles about which we geometrize do not exist in any other sense than as abstract constructions from these concrete experiences. Whosoever admits this, as even roughly and approximatively descriptive of the process of geometrizing, cannot but find the usual mathematical standpoint regarding the modern development of geometry as other than confused and vacillating, while at the same time comprehending how the mathematician has been driven to such a standpoint.

In part this state of things proceeds from drawing the line, but not drawing it firmly and consistently, between geometry and its practical applications. But it is also in part due to the tradition fixed through the unfortunate way in which Euclid

[1] *La Science et l'Hypothèse*, p. 51.

[2] Roughly, the shape of a saddle; or of a mountain pass, where the curvatures along and across the pass are opposed in sense.

[3] It is the negation of this proposition which Simson takes, as an axiom, in order to demonstrate Euclid's postulate. See his *Notes on the Elements*, p. 260 of the 25th edition (1841).

formulated his theory of parallels. His successors were not satisfied with the theory—and no doubt he himself must have been unsatisfied with it; but they never dared break loose from his authority, and recast the theory in such a way as to remedy its principal defect, viz. the complex and artificial character of the 12th Axiom, a proposition which expresses a process of thought so utterly different in kind from those to which the other fundamental propositions give expression. They admitted it as necessary, and yet not as a necessity of thought; as needing demonstration, and yet as apparently indemonstrable. Now if we admit, with Lobatschewsky and others, that Euclid's 12th Axiom is not a necessity of thought, is neither a real axiom nor a proposition derivable from real axioms; and if we understand Lobatschewsky's hypothesis as he apparently intended it to be understood, viz. as a real alternative to, and as in real contradiction with, Euclid's 12th Axiom; how does the case then stand? We are none the less obliged to admit that the results which Lobatschewsky deduces from his assumption are logical results, and this (remembering that we take Lobatschewsky to be dealing with the same geometrical entities as Euclid) is to admit that there are two possible but mutually exclusive planimetries of the surface which Euclid calls plane. Moreover, this conclusion is entirely independent of whether Lobatschewsky did or did not intend that which he appears to have intended—his work supplies us with the data for the conclusion, and we cannot but draw it whether he intended it or not. Now set against this the received view, as expressed by Stephen Smith, that Lobatschewsky's hypothesis is in effect to assume that the plane has negative curvature. It was reserved for an Italian mathematician, Beltrami—as Professor Smith goes on to say—to show that the plane geometry of Lobatschewsky is identical with the geometry of a pseudo-spherical surface, i.e. of a surface of constant negative curvature. That is to say, the metrical relations of the sides and angles of a rectilinear triangle, strictly so called, derived by Lobatschewsky from his anti-Euclidean assumption, are identical with the metrical relations of the sides and angles of a rectilinear triangle, loosely so called, on a surface of constant negative curvature.

Now if we cannot admit that the relations of the sides and angles of a geodesic triangle on a surface of zero curvature are identical with the relations of the sides and angles of a geodesic triangle on a surface of constant negative curvature, we cannot admit the hypothesis from which this identity flows. Euclid's 12th Axiom is thus a necessity of thought; that is, it is either what Cayley, in Playfair's version of it, found it to be: a self-evident proposition, not needing demonstration; or, as I have ventured to suggest, a proposition which derives from the real axioms of direction which Euclid should have given us, but did not. And, on this view, Lobatschewsky's hypothesis should not be described as in effect equivalent to the assumption that a plane has negative curvature, which is an ambiguous piece of phraseology, but as in effect equivalent to giving an extension of meaning to the term 'plane', so that it shall mean the surface of constant negative curvature as well as that of zero curvature. And such an extension of the term, even though it may tend more to confusion than to clearness of thought, cannot be condemned as purely arbitrary, for the assigned community of name does at least correspond to a property common to the two surfaces, viz. that uniformity which is defined in mathematical language as the constancy of the measure of curvature. But this extension of meaning once admitted, analogy in conception drives us a step further on the road of metaphor, and we are committed to calling the surface of constant positive measure of curvature, as well as that of constant negative measure of curvature, a plane.

If we thus extend the meaning of the term plane, we can no doubt go on to speak of Euclid's planimetry, of Lobatschewsky's planimetry, of Riemann's planimetry; but the conceptions thus denoted do not involve that of different 'systems' of geometry, the admission of any one of which excludes the others. These 'planimetries' are merely different branches of geometry. Riemann's geometry, in two dimensions, as Mr. Poincaré remarks,[1] does not differ from spherical geometry, which is a branch of 'ordinary' geometry. And here evidently the path we have been following comes to an abrupt end: we are left in contemplation of three branches of geometry, but without the faintest indication of a departure from, or modification of, the ordinary notion of space, or of an attribution of meaning to the expression 'systems' of geometry. According to the metageometer, however, we have to distinguish between (1) surfaces of constant

[1] *La Science et l'Hypothèse*, p. 55.

positive, and of constant negative, measure of curvature in 'Euclidean' space; and (2) the plane in 'elliptic' and in 'hyperbolic' space.[1] The condition of our ability to make this distinction can of course only lie in our ability to conceive these different kinds of space. It is to the explanations of this development of the notion of space that we must now turn.

[1] See 'Geometry, non-Euclidean' in the *Ency. Brit.*, vol. xxviii, p. 670.

CHAPTER XVII

POPULAR EXPOSITIONS OF METAGEOMETRY

Helmholtz's explanation of a means by which we may attain variety in the conception of Space.—The way suggested is through analogy with the differing space-conceptions of logical ' two-dimensional beings ' inhabiting different kinds of surface-worlds.—The space-conceptions of these figurative beings are, however, nothing but geometrical abstractions from our own spatial experience, clothed in allegorical language ; and the ground of the analogy is thus itself an illusion.—Lotze's ineffective attack on Metageometry.—Clifford's attempt to carry the conception of Elementary Flatness from the surface to space.—The false analogy involved in Clifford's reasoning.

LET us suppose that the phrase ' species of space and the systems of geometry relevant to these several species ' does briefly indicate, for some few human beings, a certain real process of conception, of which the great mass of human beings have no cognizance. For this great mass or—let us go so far—even for the educated and intelligent man, the expression ' species of space ' is simply devoid of meaning. In saying that there are different kinds of space, metageometers of course do not intend to assert that different kinds of space ' exist ' ; but that we are able to conceive different kinds of space, and that we do not know which of these differently conceived spaces is ' our ' space, or the space which ' exists '.

There are two ways in which the non-Euclidean geometer or, as we might perhaps call him, the more-than-Euclidean geometer, explains how we can attain to a generalized conception of space and a correspondingly generalized conception of geometry. There is the academical and mathematical way taken by Riemann in his essay *Ueber die Hypothesen, welche der Geometrie zu Grunde liegen* (1854),[1] and there is the simpler and less technical way, better suited than Riemann's to the mind not highly trained in mathematical methods of expression. The latter mode of explanation is employed by Helmholtz in his discussion of the Origin and Significance of Geometrical Axioms.[2] I do not know

[1] Not published till 1867, after Riemann's death.

[2] *Popular Lectures on Scientific Subjects*, translated by E. Atkinson, Ph.D., F.C.S., 1881.

whether he was the original inventor of it, but it has certainly met with approval, for it has been repeated with insignificant variations by almost every mathematician who has sought, in current phraseology, to make clear to an intelligent inquirer the non-Euclidean doctrine. And, as will I think be seen further on, the purely analytical (in the mathematical sense of the term) part of Riemann's exposition really depends for its geometrical relevance upon whether it is or is not a fact that the ordinary notion of space is susceptible of generalization or transformation, upon whether we can or cannot conceive space as having other attributes or ' properties ' than those we commonly assign to it.

Helmholtz's explanation is rather long, but as it has obtained wide currency, though not altogether without criticism, it will be as well to quote it *in extenso* :

' Let us, as we logically may, suppose reasoning beings of only two dimensions to live and move on the surface of some solid body. We will assume that they have not the power of perceiving anything outside this surface, but that upon it they have perceptions similar to ours. If such beings worked out a geometry, they would of course assign only two dimensions to their space. They would ascertain that a point in moving describes a line, and that a line in moving describes a surface. But they could as little represent to themselves what further spatial construction would be generated by a surface moving out of itself, as we can represent what would be generated by a solid moving out of the space we know. By the much abused expression "to represent" or "to be able to think how something happens" I understand—and I do not see how anything else can be understood by it without loss of all meaning—the power of imagining the whole series of sensible impressions that would be had in such a case. Now as no sensible impression is known relating to such an unheard-of event, as the movement to a fourth dimension would be to us, or as a movement to our third dimension would be to the inhabitants of a surface, such a "representation" is as impossible as the "representation" of colours would be to one born blind, if a description of them in general terms could be given to him.

' Our surface-beings would also be able to draw shortest lines in their superficial space. These would not necessarily be straight lines in our sense, but what are technically called *geodetic lines* of the surface on which they live ; lines such as are described by a tense thread laid along the surface, and which can slide on it freely. I will henceforth speak of such lines as the straightest lines of any particular surface or given space, so as to bring out their analogy with the straight line in a plane. I hope by this expression to make the conception more easy for the apprehension of my non-mathematical hearers without giving rise to misconception.

'Now if beings of this kind lived on an infinite plane, their geometry would be exactly the same as our planimetry. They would affirm that only one straight line is possible between two points; that through a third point lying without this line, only one line can be drawn parallel to it; that the ends of a straight line never meet though it is produced to infinity, and so on. Their space might be infinitely extended, but even if there were limits to their movement and perception, they would be able to represent to themselves a continuation beyond these limits; and thus their space would appear to them infinitely extended, just as ours does to us, although our bodies cannot leave the earth, and our sight only reaches as far as the visible fixed stars.

'But intelligent beings of the kind supposed might also live on the surface of a sphere. Their shortest or straightest line between two points would then be an arc of the great circle passing through them. Every great circle, passing through two points, is by these divided into two parts; and if they are unequal, the shorter is certainly the shortest line on the sphere between the two points, but also the other or larger arc of the same great circle is a geodetic or straightest line, i.e. every smaller part of it is the shortest line between its ends. Thus the notion of the geodetic or straightest line is not quite identical with that of the shortest line. If the two given points are the ends of a diameter of the sphere, every plane passing through this diameter cuts semicircles on the surface of the sphere, all of which are shortest lines between the ends; in which case there is an equal number of equal shortest lines between the given points. Accordingly, the axiom of there being only one shortest line between two points would not hold without a certain exception for the dwellers on a sphere.

'Of parallel lines the sphere-dwellers would know nothing. They would maintain that any two straightest lines, sufficiently produced, must finally cut not in one only, but in two points. The sum of the angles of a triangle would be always greater than two right angles, increasing as the surface of the triangle grew greater. They could thus have no conception of geometrical similarity between greater and smaller figures of the same kind, for with them a greater triangle must have different angles from a smaller one. Their space would be unlimited, but would be found to be finite or at least represented as such.

'It is clear, then, that such beings must set up a very different system of axioms from that of the inhabitants of a plane, or from ours with our space of three dimensions, though the logical powers of all were the same; nor are more examples necessary to show that geometrical axioms must vary according to the kind of space inhabited by beings whose powers of reason are quite in conformity with ours.'

The kind of space which we inhabit is, then, known to us in the same way as we may logically suppose the kind of space, inhabited by the two-dimensional beings, would be known to them, if they were confined in their movements, perceptions, and measurements to a very small part of their two-dimensional space. We may consider the two-dimensional beings as living and moving on a very small part of a surface which is either plane, or spherical, or pseudo-spherical. The conditions of their geometrical experience will be such as to furnish them with the conception of congruence, and their metrical axioms will be determined at least in so far that they must be relevant to this conception. We must not suppose that their axioms would be absolutely determinate, for this in effect would be to turn these axioms into necessities of thought, in the sense of *a priori* principles independent of the actual spatial experience of these beings. What we must suppose, as a logical development of Helmholtz's view, is that the two-dimensional beings will, in their search for indubitable geometrical axioms, eventually find themselves confronted (as is alleged to be the case with us) with at least three sets of geometrical axioms, each set being consistent with their experience, possible but not necessary.

According to this theory, the two-dimensional beings, if *per impossibile* they could become aware of the third dimension of space, and thus be able to construct the notion of a surface, would correlate the three sets of axioms respectively with the three possibilities that their space was a plane, a spherical, or a pseudo-spherical surface. If this fiction of two-dimensional beings is to be of any use to us in affording an analogy by means of which we may be put in the way of conceiving our space as one of several possible kinds of space, it is clear that what we require to know is the nature of the several conceptions of two-dimensional space which the two-dimensional being, *not* being aware of the third dimension of space, constructs. This is no doubt an easy matter for the metageometer, who, being already in possession of the notion of kinds of three-dimensional space, can judge, by analogy with his own process of thought in constructing this notion, how the two-dimensional being arrives at the analogous notion, and what the precise nature of that notion is. But for those who do not already conceive space as one of several possible kinds of space, this story about two-dimensional beings is useless, save in so far that, being supplied by the meta-geometer in elucidation of the conception in question, the fact affords evidence of some confusion of thought in the mind of the metageometer.

We need only ask ourselves a very simple question in order to discover the precise value of this analogical illustration. Admitting the fiction of two-dimensional beings living and moving in these several two-dimensional spaces, how do we go on to conceive their fictitious experiences and conclusions? To ask this question is at once to see that the whole of this ingenious story of Helmholtz's is merely an allegorical or metaphorical way of giving expression to our own geometrical abstractions, so that the end of the story leaves us precisely where we were at the beginning of it. Presented as Helmholtz presents it, that is, as a real and fecund analogy in conception, it involves the wholly inadmissible, the preposterous, metamorphosis of our spatial *abstractions* into the spatial *experiences* of the two-dimensional beings. The two-dimensional beings clearly cannot think *our* spatial abstractions, for our concrete three-dimensional experience is a necessary condition of forming these abstractions, so that the only logical, and perfectly useless, conclusion is that their experience is as inconceivable to us as ours is to them. It is only if we take Helmholtz's argument in the sense in which he obviously did not intend to confine it, viz. as merely an allegorical statement of our own spatial abstractions, that the substitution of our abstractions for the experience of the two-dimensional beings is admissible. It is a logical handling of the allegory; but, in that case, as we have seen, no conceptual development whatever is accomplished.

That Helmholtz, while believing himself to be advancing a serious philosophical and scientific argument, was really the dupe of his own fable, is confirmed by a passage which occurs a few pages further on in his lecture:

'Inhabiting, as we do, a space of three dimensions, and endowed with organs of sense for their perception, we can represent to ourselves the various cases in which beings on a surface might have to develop their perceptions of space; for we have only to limit our perceptions to a narrower field. It is easy to think away perceptions that we have; but it is very difficult to imagine perceptions to which there is nothing analogous in our experience.'

The simple fact is that we can no more think away perceptions that we have than we can imagine perceptions not in analogy with those we have; not, at all events, in the sense relevant to Helmholtz's statement. He was evidently thinking, not of disregarding perceptions that we have, but of abstracting from the three-dimensionality of space, which is involved in all perception, recollection, and imagination of the external world; and he thus turns our process of spatial abstraction or of analysing spatial experience into the experience of the two-dimensional beings. As an elucidation or vindication of the reality and significance of the non-Euclidean doctrine the argument appears to me to be from beginning to end bad philosophy and bad science.

Mr. Bertrand Russell is, so far as I know, the only writer on the subject of non-Euclidean geometry who shows a distrust of this allegorical argument; and even his condemnation is but half-hearted. He speaks of Helmholtz's romances about Flatland and Sphereland as at best only fairy tale analogies of doubtful value, and considers them to be no essential feature of Metageometry;[1] but he defends them against Lotze's attacks. They are of no value at all as exhibiting by analogy a process of conceptual development; on the other hand they are invaluable as evidence of the tendency to mysticism in the mathematician.

One important contributory cause of the hold of this so-called process of illustration on the minds of so many eminent reasoners is that it was attacked at the outset, and erroneously attacked, by other eminent reasoners. When a thinker of Lotze's eminence disputes the validity of an argument and fails to make good his case, the argument becomes thereby more authoritative: some who wavered or were sceptical become convinced. Lotze felt that Helmholtz's argument was unsound, but he failed to diagnose the nature of the unsoundness. He went the wrong way to work by disputing that the two-dimensional beings would experience and think what Helmholtz affirmed they must experience and think. He was bound to come off second-best in that argument

[1] *Foundations of Geometry*, p. 101.

if what we can affirm about the conclusions of the two dimensional beings is derived from our conception of the surface. It did not occur to him to point out that this derivation itself furnishes an immediate condemnation of the purpose to which it is put.

Now that not only Helmholtz, but other metageometers as well, have made use of this same argument, of this so-called analogy, under the impression that it helps to explain and illustrate their thesis that different kinds of space are conceivable, is, to say no more of it, evidence of some confusion of thought in the minds of these mathematicians. Mr. Russell may tell us, as he does, that these analogical fairy-tales form no essential part of metageometry—they do form part of the metageometer's stock-in-trade for the popular exposition of his subject. It is true that in the popular exposition of an abstruse subject we must expect the intricacies of it to be kept in the background, and some of its difficulties to be slurred over. But a popular illustration which, in the expounder's own mind, involves a confusion of thought of which he is himself unconscious is a very different matter, and cannot but arouse suspicion as to the genuineness or the real significance of the thesis which it is supposed to illustrate.

This suspicion would of course disappear to make room for mere mild astonishment at the vagaries of the mathematician's philosophical excursions if, apart from these vagaries, we could find any genuine exposition, popular or other, of this alleged evolution of the notion of space. It is believed, by metageometers, that such an exposition is to be found in Riemann's famous essay on the hypotheses which underlie geometry. In the next chapter we shall examine this dissertation and try to form a considered judgement as to whether that belief is well founded or not. But before we go on to this it may be worth while to ponder another popular explanation of one of the main features of the non-Euclidean doctrine, given by Clifford in his lecture on 'The Philosophy of the Pure Sciences'.[1]

One of the postulates of the science of space, Clifford tells us, is that of Elementary Flatness, or flatness in the smallest parts.[2]

[1] *Lectures and Essays*, by W. K. Clifford, vol. i, pp. 369–73.

[2] The expression is taken from Riemann's Dissertation, where it is used as equivalent to the two propositions, (1) that line-length is independent

He begins by explaining what is to be understood by this expression in relation to the line and the surface, and goes on thence to give an interpretation of it when employed in connexion with space:

'Any curved surface which is such that the more you magnify it the flatter it gets, is said to possess the property of elementary flatness. But if every succeeding power of our imaginary microscope disclosed new wrinkles and inequalities without end, then we should say that the surface did not possess the property of elementary flatness.'

This is, of course, quite clear if 'the more you magnify it the flatter it gets' is understood in the sense intended by Clifford.

'But how am I to explain'—he continues—'how solid space can have this property of elementary flatness? Shall I leave it as a mere analogy, and say that it is the same kind of property as this of the curve and surface, only in three dimensions instead of one or two?'

A mere analogy! But all explanation, even when hypothetical, consists in apprehending analogy between one set of relations and another set (compare with the notion of the relativity of knowledge); only we have first to conceive the relations which, when their analogy with others is apprehended, are thus explained or understood. By 'a mere analogy' Clifford seems to have understood 'a mere metaphor'; and a *mere* metaphor explains nothing, because it expresses no apprehended analogy.

However, Clifford thinks he 'can get a little nearer than that'; he will at all events try. He goes on to remark that if we start to go out from a point on a surface there is a certain choice of directions in which we may go. If we go on changing a direction of start, it will, after a certain amount of turning, come round into itself again. Every surface on which the amount of turning, necessary to take a direction all round into its first position, is the same for all points of the surface, is a surface which has the property of elementary flatness.

'I will now show you a surface which at one point of it has not this property. I take this circle of paper from which a sector has been cut out, and bend it round so as to join the edges; in this way I form a surface which is called a *cone*. Now on all points of this surface but one, the law of elementary flatness

of position, and (2), that the line-element is expressible as the square root of a quadric differential.

holds good. At the vertex of the cone, notwithstanding that there is an aggregate of directions in which you may start, such that by continuously changing one of them you may get it round into its original position, yet the whole amount of change necessary to effect this is not the same at the vertex as it is at any other point of the surface. And this you can see at once when I unroll it; for only part of the directions in the plane have been included in the cone. At this point of the cone, then, it does not possess the property of elementary flatness; and no amount of magnifying would ever make a cone seem flat at its vertex.'

No amount of magnifying will turn a geometrical point into part of a surface. In that respect all the points of the conical surface are alike. Elementary flatness of a surface has no meaning unless by 'at a point' we understood 'in the immediate neighbourhood of a point'. But, again, there is no difference in this respect between the vertex point and any other point of the surface. It possesses the property of elementary flatness, as Clifford at first defines it, as well in the immediate neighbourhood of the vertical point as in that of any other point. No matter what part of the surface you magnify, it becomes flatter the more you magnify it.

Still, there is the difference, by whatever name we please to call it, to which Clifford draws attention; the difference between the aggregate of directions in which we can set out on the surface from the vertex point, and this aggregate in relation to any other point contained in the surface. We shall, however, find a similar difference if, instead of the vertical point, we take any point on the circumference which limits the conical surface. The vertical point and the points on the limiting circumference, although we speak of them as points of the surface, are in fact related to the surface in a wholly different manner from the others which are contained in it. The vertical point is neither in the surface nor out of it; it is a boundary of the surface. And it is precisely with this relation that the difference to which Clifford calls attention is bound up. If, for instance, we suppose the conical surface to be 'produced' through the vertex so as to form the double conical surface, then the vertex ceases to be a boundary of the surface, and at the same time the distinction in question disappears, so far at least that we have now from the vertex, as from any other contained point, opposite directions in which we can start on the surface, and it is no longer admissible that the choice of directions is at this point necessarily less than it is at any other.

Now by whatever name we may indicate this particular difference of property between a surface such as a cone and one such as, e.g., a sphere; it is clear that what Clifford will have to accomplish is the conceiving, for space, of a difference analogous to this difference for the surface. But this difference is, in essence, as we have just seen, the difference between the presence and the absence of a boundary. Hence his attempt must necessarily fail; for whether we agree or not with Riemann that the infinity of space is not a necessary conception, we cannot but agree with him that its unboundedness is a necessary conception. Thus when Clifford goes on to explain that the difference between a space which possesses the property of elementary flatness and one which does not, is that, in the former, but not in the latter, the total amount of solid angle (or the whole aggregate of directions) is the same round every point, we see that, so far as the credit of metageometry is concerned, he would have been better advised to remain content with the 'mere analogy'.

CHAPTER XVIII

RIEMANN'S DISSERTATION ON THE FOUNDATIONS OF GEOMETRY

Riemann's initial sketch of the filiation of ideas in the Dissertation.—Analysis of the logical interconnexion of these ideas.—The notion of space as a particular case of a more general notion.—The alleged empirical nature of geometrical premisses cannot be established *a priori*.—The alternative premiss to Euclid's so-called Axiom 10, which emerges from Riemann's investigation.—The conception of a Manifold.—Obscurity of this conception as explained by Riemann.—And by Helmholtz.—Consequent indeterminateness of the relation in which this conception stands to that of space.—The conception which, under the term manifold, Riemann subsequently subjects to mathematical analysis, is a clear conception; but the conception of space is not a particular case of this conception.—The conception of the measure of curvature of a manifold.—Relevance of this conception to space.—Emptiness of the analogy from Gauss's measure of curvature of surfaces.—The non-relation between various space-constants.

THE notion of space and the first principles of construction in space are, according to Riemann,[1] assumed in geometry as things given; geometrical definitions are merely nominal, while the true 'determinations' appear in the form of axioms, or assumptions. By this we are to understand, I presume, that in geometry we assume the 'existence' of the geometrical entities defined. What relevance the mental process called 'assumption' has to the 'existence' of the geometrical entities defined is left unexplained. However, it is, according to Riemann, a consequence of this state of things that the relation of these assumptions (presumably *inter se*) remains in obscurity: we perceive neither whether nor how far their connexion is necessary, nor, *a priori*, whether this connexion is possible.

The precise way in which the obscurity which envelops the inter-relations of these assumptions results from the alleged fact that we make these assumptions is very far from clear. But this need not discourage us; for Riemann immediately proceeds

[1] Throughout this chapter I have used the translation of Riemann's dissertation contained in Clifford's *Collected Mathematical Papers*. This translation is considered, indeed, to be not wholly above reproach; but I know of no other, and in it we have at least an interpretation of Riemann by a philosopher and mathematician.

to attribute this obscurity to something else, viz. that from Euclid's time to that of Legendre the notion of multiply extended magnitudes (in which space-magnitudes are included) remained entirely uninvestigated. The first step towards dispelling this obscurity lies therefore in constructing the general notion of a multiply extended magnitude. Riemann then briefly indicates the conclusions which emerge from the construction and investigation of this general notion. They are as follows: That a multiply extended magnitude is capable of different sets of measure-relations, and consequently that space is only a particular case of a triply extended magnitude. Hence it follows as a necessary consequence that the propositions of geometry cannot be derived from general notions of magnitude, but that the properties which distinguish space from other conceivable triply extended magnitudes are only to be deduced from experience. Thus arises the problem, to discover the simplest matters of fact from which the measure-relations of space may be determined; a problem which from the nature of the case is not completely determinate, since there may be several systems of matters of fact which suffice to determine the measure-relations of space—the most important system for our present purpose being that which Euclid has laid down as a foundation. These matters of fact are, like all matters of fact, not necessary, but only of empirical certainty; they are hypotheses. We may therefore investigate their probability, which within the limits of observation is of course very great, and inquire about the validity of their extension beyond the limits of observation, on the side both of the infinitely great and of the infinitely small.

We may pause here to consider how these conclusions hang together, taking for granted the original deduction that a multiply extended magnitude is capable of different measure-relations.[1]

There is, first, the conclusion that, a multiply extended magnitude being capable of different measure-relations, it follows that space is only a particular case of a triply extended magnitude.

[1] Throughout this chapter I must beg the reader to bear in mind that the discussion proceeds from the ordinary standpoint that geometrical premisses are *assumptions*. From my own point of view (viz. that these premisses, with the exception of the Axioms of Magnitude and Direction, are definitions of geometrical abstractions, to which such notions as 'true', 'false', 'doubtful', 'certain', &c., are irrelevant) many of the arguments used would be meaningless.

As a deduction this is somewhat puzzling. For if the general notion of a multiply extended magnitude is constructed—as subsequently appears to be the case—from particular notions of which space is one, then evidently space will be a particular case of the general notion of multiply extended magnitude in general, whether this is or is not capable of different measure-relations. On the other hand, if the general notion is not thus constructed, it does not seem possible that space can be a particular case of it; for *ex hypothesi* space is then not one of the particulars in relation to which the said notion is general. There is here apparently some confusion; I think it will be seen to be real when we come to consider the construction of the general notion.

Next, admitting as a datum that space is only a particular case of the general manifold or multiply extended magnitude, we have the necessary consequence that the properties of space cannot be deduced from general notions of magnitude, but can be determined only by experience.[1] This argument seems to involve a *petitio principii*, or something very like one. For the conclusion that the premisses of geometry can only be empirical depends, in Riemann's argument, upon the tacit assumption that they are not *a priori*, i.e., that they are empirical, since the only source of knowledge which we can set up in contradistinction with the empirical is the *a priori*. The argument is, in short, this: The premisses of geometry are either *a priori* or empirical; they cannot be *a priori* because they cannot be derived from general notions of magnitude—clearly an argument which assumes that the domain of the *a priori* in mathematics is confined to general notions of magnitude, and which thus, by assumption, affirms the premisses of geometry to be empirical. The fact is that if the premisses of geometry are empirical, this does not follow from Riemann's argument, but from the fact, if a fact it is, that these premisses, or some of them, are found not to possess 'apodeictic' certainty.

We pass from this point, again accepting merely as a datum that the premisses of geometry and the measure-relations involved in these premisses are only empirically determinable. We then

[1] That is to say, all the sets of measure-relations conceivable of the general manifold are *possible* in relation to the particular manifold space, but which of these possible sets is the actual set is a matter which can only be determined empirically.

have the problem which Riemann so oddly expresses. Matters of fact are not usually regarded as hypotheses; on the contrary, we commonly contrast matter of fact with hypothesis. But one sees clearly enough what Riemann means. By matters of fact he intends us here, I suppose, to understand generalizations from experience, all of which are, from the limitations of experience, hypothetical. To put it briefly, the Euclidean premisses are (according to Riemann) generalizations from experience, and as such possess no more than empirical validity. Whether Riemann regarded this as a conclusion which follows necessarily from his investigation is not by any means clear, although, as we have seen, he appears to claim that it does. At all events it does not seem possible to admit it.

But the most interesting feature of Riemann's dissertation is that, as one of the results of the investigation of the manifold in general, and of its conceivable measure-relations, he is led to the conclusion that we must extend the domain of possibility with regard to the measure-relations of space beyond the limits contemplated by his revolutionary predecessors Gauss, Lobatschewsky, and Bolyai. For some two thousand years it had been admitted that Euclid's Axiom of Parallels was at all events theoretically an unsatisfactory premiss. It was not 'self-evident', and the numberless attempts to derive it from the other and unquestioned premisses had one and all been failures. The investigations of Lobatschewsky and Bolyai went to show that these attempts must remain fruitless. In effect Riemann now discovers that there is at least one of these hitherto unquestioned premisses of Euclid which must no longer remain unquestioned, but must enter into the same category with the axiom of parallels, viz. Axiom 10, that two straight lines cannot intersect twice; or, what comes in the end to the same thing, that not only are there the two alternatives maintained by Lobatschewsky, i.e. that through a point situated outside a given straight line there must be either only one parallel to the straight line, or an infinite number of parallels;[1] but that a third possibility must be admitted: there may be no parallel to the given line. These three possibilities Riemann connects with a certain quantity which he calls the 'measure of curvature' of a manifold.

[1] Parallel in the sense of non-secant. Lobatschewsky defines the parallel as the limit between the secants and the non-secants.

In the case of the particular manifold space, Euclid's assumption corresponds with the value of this measure being equal to zero, Lobatschewsky's alternative assumption with this measure having a negative value, and the third possibility with this measure having a positive value.

Thus, as a result of this investigation, we find ourselves driven to admit as possible that which we had previously always regarded as self-evidently impossible. But Riemann's conclusion can itself ultimately rest upon nothing more secure than the self-evident. If, then, we are to admit Riemann's conclusion, we have to distinguish between the self-evident and the self-evident; between the self-evident *a priori* or apodeictically certain and the empirically self-evident or non-apodeictically certain. The latter is open to the possibility of doubt, the former not. But this distinction was common ground to thinkers long before the time of Riemann's dissertation, and geometrical axioms (with the exception of the axiom of parallels) were commonly adduced as instances of the self-evident *a priori*. To admit Riemann's conclusion is thus to admit doubt into the premisses upon which he founds it, for if we may have been mistaken (for a brief period of 2,000 years) as to the nature of the premisses from which Euclid argues, so may we also be mistaken as to the nature of the premisses from which Riemann argues.

But does Riemann's dissertation really leave us in this quandary? We have already seen, from a brief criticism of his own preliminary account of the filiation of his argument, that there is good reason for questioning, if not rejecting outright, the supposition which he apparently entertained that the empirical nature of the premisses of geometry is established as a necessary consequence of his investigation. If this is indeed a delusion, it is fairly plain that the origin of it must lie in some confusion of thought respecting the relation in which the notion of space stands to the general notion which is the subject of his investigation: the general notion of an *n*-fold extended magnitude.

Now what, exactly, is this general notion; and further, in what sense can the notion of space be included in, or be a particular case of, this general notion? These questions are not easy to answer, because Riemann's own explanations are obscure. This is no mere *ipse dixit* of mine; it has been admitted by authorities who have nevertheless accepted Riemann's conclusions;[1] but these authorities do not appear to have reflected that obscurity of the general notion involves obscurity of the precise relation between it and the notion of space. And yet, upon the precise nature of this relation depends the import of Riemann's dissertation, since the avowed object of it is to dispel the obscurity, in which the properties of space are involved, by means of the construction and investigation of this general notion. Whether the expression 'an *n*-fold extended magnitude' corresponded in Riemann's mind to a definite and stable conception can, however, only be surmised from the explanations he gives of it. That he was himself conscious of some obscurity in expression, if not in idea, may be not unfairly inferred from the fact that he prefaces the explanation itself with a plea for indulgent criticism on the ground of his lack of practice 'in such undertakings of a philosophical nature where the difficulty lies more in the notions themselves than in the construction'—a statement which, although the motive which prompts it is clear, is itself so obscure in expression as to raise painful apprehension as to the intelligibility of what is to follow.

This apprehension is intensified by the very first step with which Riemann starts his explanation. 'Magnitude-notions,' he says, 'are only possible where there is an antecedent general notion which admits of different specializations.' This is not illuminating as an introduction. Were it not for the obvious sincerity of the writer, one might be tempted to ask whether this were not an instance in verification of Talleyrand's sardonic remark that language was given to man to enable him to disguise his thoughts. Are the 'specializations' of the antecedent general notion the particular notions subsumed under the general notion, i.e., are they the magnitude-notions in question? If so, the antecedent general notion is none other than the notion of magnitude in general; but in that case the statement is disputable, and, in my opinion, altogether erroneous. A general notion is not antecedent to the particular notions subsumed under it. The particular and the general condition one another. A notion cannot be general (or particular) otherwise than in relation to a notion which is particular (or general). Thus if Riemann meant that the notion of space as a magnitude of

[1] See Mr. Russell's *Foundations of Geometry*, pp. 66–8, also p. 15 and the footnote relative to Veronese.

a particular kind is possible only if we have an antecedent notion of magnitude in general, he started his investigation with an inadequate conception of our mental processes.[1]

'According as there exists among these specializations a continuous path from one to another or not they form a continuous or discrete manifoldness.' Notions whose specializations form a discrete manifoldness are extremely common. 'On the other hand, so few and far between are the occasions for forming notions whose specializations make up a continuous manifoldness, that the only simple notions whose specializations form a multiply extended manifoldness are the positions of perceived objects and colours.' The notion of a discrete manifoldness would thus appear to be the notion of an aggregate in general. But it is with manifoldness as continuous that Riemann is more particularly concerned, and here he gives us definite simple instances of a continuous manifold, viz. the positions of perceived objects and colours.

We remark that the manifoldness of the beginning of the paragraph has become at the end of it an 'extended' manifoldness. Are the positions of perceived objects extended? But presently this is silently dropped and space is substituted for the positions of perceived objects. Is space extended? It is the condition of anything being extended. The positions of perceived objects thus become specializations of space, and form the general notion of space. They should thus be particular notions of space, and this appears to have no meaning at all. Colours are specializations of the manifold of colours; but what are we to understand here by the *manifold* being 'extended'?

'Definite portions of a manifoldness, distinguished by a mark or a boundary, are called Quanta. Their comparison with regard to quantity is accomplished in the case of discrete magnitudes by counting. In the case of continuous magnitudes by measuring. Measure consists in the superposition of the magnitudes to be compared; it therefore requires a means of using one magnitude as standard for another.'

A quantum of space, understanding by this the content of

[1] It is also a question whether, in a fundamental investigation such as this of Riemann's, we are not liable to fall into confusion through speaking of space as a magnitude; for when we speak of parts of space, and thus nominally turn space into a magnitude, what we really have in mind is magnitude-relation of figures.

a figure, is intelligible. What is a quantum of the colour-manifold? What is the relevance of measure to this manifold (not to the colours which make it up), and what, in the name of common sense—I apologize for alluding to it in a subject which so evidently transcends it—has superposition to do with the case of the colour-manifold? It is not surprising that Riemann withheld his essay from publication from the year 1854, when it was read to a small circle, till his untimely death in 1867, owing to his wish to make certain changes in it.

Helmholtz's name being closely associated with Riemann's in connexion with the pangeometrical theory, we naturally turn to his works for an elucidation of Riemann's conception. The fullest explanation which he gives of it is to be found in his lecture 'On the Origin and Significance of Geometrical Axioms.'[1] It is as follows:

'The number of measurements necessary to give the position of a point is equal to the number of dimensions of the space in question. In a line, the distance from one fixed point is sufficient, that is to say one quantity; in a surface, the distances from two fixed points must be given; in space, the distances from three; . . . or, as is usual in analytical geometry, the distances from three co-ordinate planes. Riemann calls a system of differences in which one thing can be determined by *n* measurements an "*n*-fold extended aggregate" or an "aggregate of *n* dimensions". Thus the space in which we live is a three-fold, a surface is a two-fold, and a line is a simple extended aggregate of points. Time is also an aggregate of one dimension. The system of colours is an aggregate of three dimensions, inasmuch as each colour, according to the investigations of Thomas Young and of Clerk Maxwell, may be represented as a mixture of three primary colours, taken in three different quantities. . . . In the same way we may consider the system of simple tones as an aggregate of two dimensions, if we distinguish only pitch and intensity, and leave out of account differences of timbre.'

We notice that there is here precisely the same ambiguity in carrying the term 'extended' to the general notion as in Riemann's own explanation. Indeed, in this passage from Helmholtz our attention is naturally drawn to it; for it will be observed that while he speaks of space, the surface, and the line as extended manifolds, he avoids the use of the term in

[1] *Popular Lectures on Scientific Subjects*, second series, pp. 45, 46. See also his *Wissenschaftliche Abhandlungen*, Bd. ii: 'Ueber die thatsächlichen Grundlagen der Geometrie.'

connexion with the manifolds of moments, colours, and simple tones. Yet the general notion we are supposed to form is one under which we must be able to subsume all these particular notions, and it is at the same time to be one which involves that of extension.

This inconsistency or internal contradiction manifests itself in other ways from the moment we analyse the explanation at all rigorously. The general notion which, according to Helmholtz, we have to construct is that of 'a system of differences' in general. The expression is vague, but this may be inevitable, and we are given plenty of examples. It must be remarked, however, that if we are to form this notion, general in relation to the particular instances given, the term 'differences' must have an unambiguous sense predicable in turn of all the particulars. So far is this from being the case that we find the term 'differences' stretched so as to cover notions which exclude one another (just as we have seen in the case of the term 'extended'). Two colours in the colour-manifold, two tones in the tone-manifold, differ; but in that sense of the term 'difference' two points in space do not differ, they are identical; their positions, in other words their co-ordinates, differ. *Pairs* of points differ—in the relation of apartness between the individuals of the pairs; that is, distances may differ. But a distance is not an element of the manifold in the sense required for analogy with the other manifolds; it is not determined by three measurements as a colour is thus determined in the colour-manifold.

Now it has frequently been remarked, in relation to Riemann's dissertation, that the notion of distance is not relevant to some of these manifolds; that, indeed, it is characteristic of space by contrast with other three-dimensional manifolds; but it has not been noticed that with this fact is connected an internal contradiction which manifests itself in the ambiguous use of the terms 'extended' and 'differences' in Riemann's and Helmholtz's explanations, and that we cannot in reality form a general notion answering to such a description as that given of 'a multiply extended magnitude', which shall include the notion of space as a particular case of it. It is not merely that this general notion is obscure; it is, properly speaking, not a general notion at all, but an illogical jumble of notions.

Now naturally enough, if this is the case, the objection will be made, or the question raised: how is it possible for Riemann or any one else to subject a bogus general notion to mathematical analysis? The answer is, I believe, simply this: that the general notion which Riemann does subject to mathematical analysis is not the illogical jumble of notions which he calls the general notion of a multiply extended magnitude, but is a purely algebraic conception, completely independent of the explanations antecedently given and involving nothing but purely quantitative notions. It is this n-dimensional general numerical manifold whose 'measure-relations' Riemann investigates. Space is not a particular case of this n-dimensional general numerical manifold; the particular cases of the latter are all those concrete numerical n-dimensional manifolds which are subsumable under it; that is, which are implied in the generality of the algebraical expressions employed. But it of course follows that the measure-relations of space must be identical with those of some one or other of these particular manifolds. The further conclusion, however, that we must admit, for space, all the possibilities of measure-relation established by the investigation of the algebraic manifold, is one which clearly we ought not to allow unless we are prepared to grant the assumption which implicitly underlies it, viz. that Euclid's purely geometrical premisses are hypothetical. This—the hypothetical nature of purely geometrical premisses—as we have already seen in discussing Riemann's introductory sketch of the dissertation, does not *follow* from his argument; it is a condition precedent to admitting the validity of his application to space of the results of his investigation of the so-called multiply extended magnitude in general. Thus if it is a question whether purely geometrical premisses are hypothetical or not—a question which I believe to be devoid of meaning—Riemann leaves this question exactly where he found it.

But let us put it that Euclid's purely geometrical premisses are generalizations from experience and not, as I hold most of them to be, abstractions from experience about which it is meaningless to ask whether they are true or not; then, these generalizations not being necessary truths, i.e. the 'existence' of the defined geometrical entities being a matter of assumption, it will follow that there will be a possibility of setting other geometrical entities in their place—in other words, of conceiving different kinds of space from that which we commonly conceive. This, too, is involved in, or is another aspect of, Riemann's dissertation. If Riemann does show us that we can conceive different kinds of space, the case for pangeometry is made out, and we must conclude that the normal mental attitude in relation to Euclid's premisses is the result of ingrained habit, of intellectual inertia.

The question, raised in this form, brings into special prominence the second of Riemann's notions, that of the 'measure of curvature' of a manifold; for it is in the subsequent application of this notion to space that we must be led, if led we are to be, to conceive different kinds of space. Riemann gives this name 'measure of curvature' to a certain algebraic expression which is obtained from the general algebraic expression for the infinitesimal 'difference' in an n-dimensional manifold. To the continuous series of values which this quantity (i. e. the measure of curvature) may have, corresponds the continuous series of particular cases of the n-dimensional manifold; and, given the value of this quantity, the measure-relations of the manifold are determined. How does Riemann come to use such a term as 'curvature', such an expression as 'measure of curvature', in connexion with this abstract analytic conception, manifoldness of n dimensions? Partly, no doubt, because in the course of his investigation he has already been constrained to employ geometrical terms, such as point, line, direction, distance, &c., which must not be taken in their literal sense; but principally because the algebraic expression in question is analogous to the algebraic expression for the measure of curvature of a surface at any point of the surface, which was derived by Gauss from the algebraical expression for the infinitesimal distance at that point,[1] and shown by him to be equivalent to the expression $1/R_1R_2$, where R_1, R_2 are the greatest and least radii of curvature at the point.

The measure of curvature of a manifold is thus an expression in which the term 'curvature' is used metaphorically. Now, to follow Riemann, the notion of space is included as a particular case in the general notion of an n-dimensional manifold. Accordingly, for space, the expression 'measure of curvature' has only

[1] That is: $ds^2 = Edu^2 + 2\,Fdu\,dv + Gdv^2$; where u, v are the co-ordinates of the point, and E, F, G functions of u, v.

a metaphorical significance; it does not imply curvature of space in the literal sense of the term curvature. Helmholtz and other authorities are careful to warn us against the misconception that the expression 'measure of curvature of space' involves any suggestion of relations relevant to sense-perception; the expression is not to be taken literally.[1] On the other hand, when, in mathematics, we establish a general theorem, we can, as it is said, 'apply' the result in particular cases; which means that we are dispensed from repeating the same process of reasoning for any particular case. The process of reasoning in the general case, however, would have no meaning at all if we could not repeat that process in a particular case. Now if space is a particular case of the n-dimensional manifold, we must be able not only to apply the results of the investigation to space, we must also be able to follow the process of reasoning in the particular case. But here, when we attempt to do so, all the geometrical terms metaphorically employed in the general investigation necessarily recover their literal meanings, and, this being so, the term curvature, predicated of space, should have a literal meaning, unless it has no meaning at all. Well, as we have seen, it has, in this connexion, no literal meaning. The fact is, once more, that the notion of space is not, in the strict sense required, a particular case of the notion of the n-dimensional manifold; it involves more than mere notions of quantitative relation. What really happens, when we endeavour to carry over to space the ratiocinative process employed for the manifold of n-dimensions, is that we cannot imagine the constructions corresponding to the supposed non-Euclidean varieties of space. This does not trouble the metageometer; he is, in general, ready to admit it;[2] but he affirms that though we cannot *imagine*, we can *conceive* them. What we do conceive, I submit, are not varieties of space or systems of geometry, but the varieties of Riemann's n-dimensional manifold, and these are purely analytic constructions involving nothing beyond mere notions of quantitative relation.

That curvature of space is not a meaningless phrase is of course

[1] *Popular Lectures on Scientific Subjects*, series 2, p. 47.

[2] Not always. Helmholtz insisted that non-Euclidean spaces *are* imaginable (see his *Origin and Significance of Geometrical Axioms*). What he described as imaginable is—I agree with him against his critics—truly imaginable; but it is so just because it involves no modification of the ordinary notion of space.

a postulate necessary to the metageometer; without it the doctrine vanishes in a mist of words and algebraic symbols. Only there is here a difference of opinion among metageometers themselves. Some of them (the late well-known astronomer, Simon Newcomb, was one) suppose that a three-dimensional 'curved' space is curved in a four-dimensional 'plane' space. Whence does such a supposition arise? To all appearance from the impossibility, for them, of attaching any meaning to curvature of space save 'by analogy' with curvature of surface. Thus easily do we conceive four-dimensional space. On the other hand there are metageometers who will by no means admit the necessity of this supposition. 'Indeed, it is the very absence of reference to a higher Euclidean space which is chiefly interesting about the non-Euclidean spaces.'[1] Space, according to the latter authorities, can very well be conceived to have a measure of curvature which involves no reference to any higher dimensions. Mr. Russell gives a very ingenious explanation of how this is to be done;[2] only he proceeds on the supposition, 'which may be allowed to stand for the moment,' that space is a particular case of Riemann's manifold; he does not consider what is to happen when, after having allowed the supposition to stand for a moment, we have to reject it as invalid.[3] But apart from this, the explanation appears to me to involve a fallacy which touches the foundation of the metageometrical doctrine, and it will therefore be time well spent to consider it as briefly as may be consistent with clearness.

The argument is that Gauss's researches on the curvature of surfaces puts us in possession of a conception of measure of surface curvature which involves no reference to anything outside the surface: measure of curvature from *within*, as Mr. Russell calls it; that we thus see how an analogous conception for space might be formed; and that Riemann's dissertation actually constructs this conception for us.

[1] *Ency. Brit.*, vol. xxviii, p. 670.

[2] *Foundations of Geometry*, pp. 19–21.

[3] The supposition is admitted on the ground that the 'analytical conception of space' with which Riemann deals is a particular case of the continuous manifold. But if the 'analytical conception of space' is not a question-begging phrase, then the conception of a continuous manifold really does include space as a particular case, and there ought to be no need for the reservation which Mr. Russell makes.

The simplest expression for the measure of curvature of a surface at any point of the surface involves reference to the radii of curvature of the surface at the point. But, as already mentioned, Gauss obtained a formula, for the infinitesimal distance at any point of a surface, which contains no terms implying measure of anything not in the surface; and from this formula alone he constructs an expression for the measure of curvature of the surface at the point. The measure of curvature of a surface thus appears as an inherent or intrinsic property of the surface or—as Klein puts it—of the length of the line-element, and no reference to space 'outside' the surface is involved in it.

This, indeed, is a marvellous result when we reflect that Gauss demonstrates that this expression for the measure of curvature is equivalent to the previous expression $1/R_1R_2$; a demonstration which necessarily implies identity of the unit of length on and off the surface. The truth is that on a surface of *variable* curvature there can be no unit of length *intrinsic* to the surface. Mr. Russell himself remarks that 'it is logically impossible to set up a precise co-ordinate system, in which the co-ordinates represent spatial magnitudes, without the axiom of Free Mobility', and this involves constancy of the measure of surface curvature. Thus Gauss's expression for the measure of curvature 'will only be *really* free from reference to the third dimension when we are dealing with a surface of constant measure of curvature'. But we cannot 'deal with', i.e. conceive, merely one surface of invariable curvature: this would be to deprive the term 'measure' itself of any meaning in relation to surface curvature. Nor is measure of curvature conceivable as constant unless it is also conceived as variable. And, so far as the argument is concerned, it makes not the slightest difference whether we have in mind variability of curvature from point to point of the same surface, or variety of surfaces of different, but each of constant, curvature. Gauss's formula for the measure of curvature at *any* point of a surface whose curvature varies continuously from point to point, is the formula for the measure of curvature of *any* surface whose curvature is constant; and just as in the former case there is no unit of length intrinsic to the surface, so in the latter is there no unit of length common to all the surfaces, other than a unit not intrinsic to any one of them. It is therefore as impossible in the one case as it is in the other to be really rid of implicit reference to the third dimension. Thus when 'by analogy' we come to consider 'measure of curvature of space'—a phrase which has somehow to be accepted as really significant if the whole theory is not to lapse into the inane—it is no wonder that some metageometers are driven to suppose a space of higher dimensions in which 'our' space may be curved.

It is interesting to remark that this fallacy concerning measure of surface curvature which Mr. Russell appears to have overlooked in his work on the Foundations of Geometry confronts him under a wider aspect in the article on Non-Euclidean Geometry in the *Encyclopaedia Britannica.* In that article he sees with perfect clearness that if the measure of curvature or constant of a space is an inherent property of the space, there can be no quantitative relation between the constants of different spaces, the question thus arising, What is the relation between different space-constants? It is a tight corner. Mr. Russell admits the question to be 'somewhat perplexing'. To find an issue from a blind alley without retracing the steps that have led us into it certainly is perplexing. The tentative answer which Mr. Russell gives to this question appears to me to be no less perplexing than the question itself. Yet the answer is in a sense easy, a foregone conclusion. Since space-constants can have no quantitative relation, they can evidently only differ qualitatively—like the shades of red, as Mr. Russell suggests.

It is a simple answer, but a solution of the question it is not; for to accept it is to make chaos of the relation of the general to the particular. The general notion of a space-constant is, according to this view, purely a qualitative notion. The particulars subsumed under it, i.e. the different space-constants, cannot therefore be other than purely qualitative. Yet they are also necessarily quantitative notions, since the space-constant of any non-Euclidean space has a particular length, that is, has some quantitative relation to a selected unit of length in that space. Thus these particular notions are at once subsumable and not subsumable under the general notion. When we can consider any particular thing as qualitatively or quantitatively differing from any others, the general notion which comprises the particulars must also have a qualitative and a quantitative aspect.

Take, for instance, the case of red and the shades of red. The relation of general to particular is here perfectly clear, no matter whether we regard red as a common or general quality and its shades as particular cases of the quality, or whether we regard the shades as the wave-lengths x_1, x_2, x_3 . . . x_n, and red as the series x of wave-lengths, where x may have any of these particular values. The explanation, in short, involves our forming a conception under conditions which do not admit of our forming one. Would it not be more rational to retrace our steps?

Thus if we had to choose between the metageometers who see in the measure of curvature of space an inherent property of space and those who see in it a relation to a space of higher dimensions, it would be simpler to take our stand with the latter, even though we should thus, according to Mr. Russell, lose the chief interest of non-Euclidean spaces. A four-dimensional space has at least the great advantage that it needs no explanation. You must simply accept the phrase either as expressing a real conception or as evidence of a mystical delusion.

CHAPTER XIX

CAYLEY'S THEORY OF DISTANCE. GEOMETRY AND MENSURATION

Cayley's Theory, though non-Euclidean, does not imply non-Euclidean spaces.—It is in expression a Theory of Distance, but in conception it is a theory of different systems of descriptive equivalence, depending upon the choice of the fundamental figure which Cayley calls the Absolute.

Astronomical observations as a test of geometrical theory.—Illusiveness of this test.—Meaning of the rectilinear propagation of light-rays.—Interpretation, from the Euclidean and from the non-Euclidean standpoints, of an assumed non-Euclidean result of measurement.—The ordinary notion of Direction has no logical foothold in non-Euclidean space-conceptions; it is nevertheless employed in the alleged construction of these space-conceptions.

CAYLEY'S views on fundamental questions connected with geometry have been cited on several occasions in the course of this work. We may not always agree with them, but, with one exception, they are at least readily comprehensible. The exception is the standpoint from which he regards the axiom of parallels. When one becomes acquainted with the nature of his work on non-Euclidean systems of geometry however, this standpoint, though perhaps not as perfectly clear as one could wish it to be, at least grows more intelligible.

Cayley may be regarded as the founder of what Klein has called the 'third period' in the development of non-Euclidean geometry; Gauss, Lobatschewsky, and Bolyai being the principal figures in the first period, Riemann and Helmholtz in the second. In the first period the foundation of the debate is plainly the alleged, and generally acknowledged, incertitude of Euclid's axiom of parallels; this and nothing more: are there many parallels, through a given point, to a given straight line, or is there one only? The alternative system is developed simply and unambiguously as a logical consequence of admitting the first of the two hypotheses; and the possibility of this development is held to show the indemonstrability of the so-called 12th Axiom from the other indubitable premisses of the elements. There is no question, there is at least no overt question, of any possible modification or generalization of the notion of space. In the second period, a proposition which had always been regarded as indubitable, Euclid's so-called Axiom 10, is for the first time called in question; and at the same time there emerges the doctrine of different kinds of space; a doctrine which obviously reacted upon and modified the point of view from which the work of the first period had been generally regarded.

With Cayley, the metageometrical doctrine undergoes a complete transformation. Or, again, we may say that there are now two doctrines. The supposition that we can conceive different kinds of space is not rebutted, it is ignored. We continue, however, to have different systems of geometry, and nominally the same systems as in the other doctrine. This is accomplished by conceiving different kinds of distance instead of different kinds of space. Since, according to Cayley, we can have either the Euclidean, or any one of the non-Euclidean, systems of metrical relations in the Euclidean plane, his attitude as a metageometer who admits the self-evidence of Euclid's postulate becomes easier to understand.

I do not propose to say very much about this, the latest phase of non-Euclidean geometry, because with the disappearance, or the putting aside, of the supposition that we are able to conceive different kinds of space, there disappears also one source of mystical illusion. It is true that in the damming of this source another is opened up, viz. that of the analytical and geometrical imaginaries; for Cayley's theory involves their use. But we are already familiar with this source of illusory judgements, and I have no intention of discussing it over again in relation to this theory. Nevertheless, in the supposition which underlies the doctrine, the supposition that we can conceive different kinds of distance, or different forms of metrical equality (if to 'conceive' is the right word, for Cayley, whether by accident or by design, avoids the term), there lurks an illusion, though of a kind somewhat different from that contained in the supposititious conceiving of kinds of space.

Cayley's metageometry is contained in his Sixth Memoir upon Quantics;[1] it is a 'Theory of Distance', developed both from the analytical and the purely geometrical points of view. His main object, as he himself explains in the first paragraph of the

[1] *Collected Mathematical Papers*, vol. ii.

Memoir, is to establish the notion of distance upon purely descriptive principles. What are purely descriptive principles in geometry? They are commonly understood to refer to qualitative relations, to exclude all reference to metrical relations. The investigation thus appears to start from, or to be founded upon, a contradiction. To establish the notion of distance upon principles which exclude this notion seems at first sight to be an absurdity. But it is only the phraseology which is absurd, because it does not express, in accordance with the usual conventions of language, the actual process of thought. The result of this violation of the conventions of expression is ambiguity in the doctrine itself. Hence the conflicting opinions which have arisen as to the significance of the theory.

There are no miracles in science, and Cayley does not perform this of establishing the notion of distance upon purely descriptive principles. What he does is, in substance, to use the expression 'metrical equality' to denote certain descriptive equivalences. But these equivalences remain descriptive or projective, even though we please to call them equal distances;[1] the notion is not changed because we give it the name of another notion. Professor Klein, in his lectures on Non-Euclidean Geometry, explains Cayley's theory of distance, and shows the nominal connexion of that theory with that of non-Euclidean geometry, i.e. metrical geometry in relation to different conceptions of space; but he heads this chapter (chap. ii, part i) 'Cayley's Maassgeometrie', Cayley's Metrical Geometry—and in this, of course, he does no more than follow Cayley's own phraseology; but it tends to promote the ambiguity. The connexion between the theory of distance and the metageometrical doctrine of the second period which Klein works out in detail, leaves the use of such terms as 'metrical', 'distance', 'equality', in relation to Cayley's geometry no less arbitrary than before. The connexion, as I put it above, is nominal. We have the same phraseology, the same mathematical expressions. We are said to have the same results: we do, indeed, get the same algebraic formulae, but the geometrical

[1] See paragraphs 209, 210 of the Memoir. In 210 we get a definition of 'equidistance' from the purely geometrical (non-analytical) standpoint. Cayley gives a simple, easily understood sketch of his theory in the Presidential Address to the British Association. See vol. xi of the *Collected Mathematical Papers*, pp. 435 et seq.

relevance of the symbols of number or quantity is totally different in the two cases. We have 'co-ordinates' in the one doctrine as in the other; but while these in the earlier doctrine indicate metrical quantities, in the later they are merely numbers assigned to points, not arbitrarily, as is sometimes said, but systematically; the system being based on the projective principle involved in von Staudt's quadrilateral construction. This artifice was not contemplated by Cayley himself, but is due to Klein.[1] It gives a definite meaning to the term 'co-ordinates', which term, in Cayley's Memoir, was employed in a sense undetermined by the author himself.[2] But this ingenious invention of Klein's, and the very precision of meaning given thereby to the term 'co-ordinates', conclusively show that, in Cayley's Theory of Distance, we are not dealing with metrical notions, but only applying the names of metrical relations to notions which are not metrical but descriptive.[3] It is thus also merely the natural result of an arbitrary and ambiguous use of terms that Cayley's Theory has come to be regarded as the expression of a generalized notion of distance, including, as a particular case, the ordinary notion of distance. It might be not impossible to admit this, provided the definition of an harmonic ratio is *metrical*; but then the 'equidistant points' of the Memoir would involve contradictory notions. It is a condition of this so-called Theory of Distance retaining its significance, that 'the anharmonic ratio of four points' is a phrase expressing a descriptive property, i.e. one which does *not* involve the ordinary notion of distance; and then plainly the ordinary notion of distance cannot be a particular case of the notion to which Cayley gives the name 'distance'.

It may be remarked, in conclusion, that if Cayley's Theory of Distance is accepted for what it purports to be, that is, a generalized *metrical* geometry, then on the principle (whatever it may be) in virtue of which it is said that Euclid assumes the existence

[1] See *Nicht-Euklidische Geometrie*, pp. 338 ff.

[2] See his Note to the Memoir, same volume, p. 604.

[3] In all this, or in most of it, I seem to be in substantial agreement with Mr. Russell's views as expressed in his *Foundations of Geometry*, pp. 32–8. As he remarks (footnote, p. 35): '... the reduction of metrical to projective properties, even when, as in hyperbolic Geometry, the Absolute is real, is only apparent, and has a merely technical validity.' Rightly considered, i.e. when we look to the notions expressed and not to the arbitrary mode of expressing them, it has *no* validity.

of the straight line and plane as conceived by him, we must now say as an alternative (or as an addition ?) that Euclid assumes the existence of distance as conceived by him. The one assertion is neither more nor less significant than the other.

We pass, by a natural transition of thought, to the relation or connexion between metrical geometry—in the unambiguous sense of ' metrical '—and mensuration. Whatever views we may hold as to the so-called ' assumptions ' upon which geometry is founded, there cannot be, it would seem, much room for doubt as to the reality of assumptions in mensuration. It might even be urged that the distinction between geometry and mensuration is more nominal than real, and that thus the incertitude of the fundamental propositions of geometry is not really, but only nominally, removed by this distinction. It may at least be contended that we have to apply geometrical conclusions, founded upon the Euclidean abstractions, to problems in mensuration ; and consequently that in relation, say, to astronomical mensuration, such questions as whether light-rays are or are not straight do arise, so that the application of Euclid's geometrical deductions to these cases does involve the assumption that light is propagated in those lines which he calls straight.

That we assume something which is conveniently called the ' rectilinear propagation of light ' in applying Euclid's geometry to astronomical mensuration is not open to doubt. But what the exact nature of this assumption is, and in what sense it may be said to be the assumption of a geometrical principle, are matters which need some discussion. I postpone it in order first to consider the metageometrical standpoint, how the metageometer envisages mensuration in relation to geometry.

There are from this point of view, as we know, three principal systems of geometry, which may conveniently be called Euclid's, Lobatschewsky's, and Riemann's,[1] corresponding to three principal notions of space. We do not know which of these systems applies, absolutely, to space ; that is, we do not know which of the three notions of space is, objectively, the true notion. We

[1] Strictly speaking, it is claimed that there are four principal systems, ' elliptic ' geometry comprising two kinds, (1) where two ' straight ' lines intersect in two points, (2) where two ' straight ' lines intersect in one point only.

know that Euclid's geometrical deductions are empirically demonstrable, that his system is applicable within the very narrow limits of experience. As Lobatschewsky remarks at the end of his *Geometrical Researches on the Theory of Parallels*, 'There are no means other than those afforded by astronomical observations of testing the correctness of the results to which ordinary geometry leads. The degree of their correctness is very great, as I have shown in one of my Memoirs. Thus, in the triangles which we are able to measure, it has not yet been found that the sum of the three angles differs by as much as the hundredth part of a second from two right angles.' In substance the same view was expressed by Gauss; it is likewise implied in Riemann's dissertation, and is repeated by F. Klein in his *Lectures on Non-Euclidean Geometry*.

Lotze, as we have already seen, was a determined opponent of the metageometrical doctrine; but he seems hardly to have given sufficient thought to the subject, and his opposition was not very effective. On this particular point, viz. the relevance of astronomical observations as a test of the objective validity respectively of the several systems of geometry, he advances an argument which, though inadequate to the purpose he seems to have had in view in urging it—the vindication of Euclidean geometry as apodeictic—is nevertheless a good argument as against the above view of Lobatschewsky, Riemann, Professor Klein, and metageometers in general. The argument, which I give in the words of Mr. Russell,[1] who criticizes it, is that 'measurements of stellar triangles, and all similar attempts at an empirical determination of the space-constant are beside the mark; for any observed departure from two right angles, or any finite annual parallax for distant stars, would be attributed to some new kind of refraction, or, as in the case of aberration, to some other physical cause, and never to the geometrical nature of space.' This argument, according to Mr. Russell, is a great favourite with opponents of metageometry; it is, in his opinion, a strong argument for the empirical validity of Euclid, but as an argument for its apodeictic certainty it has an opposite tendency. The empirical validity of Euclid would seem to be a matter for observation rather than for argument. But the point which

[1] *Foundations of Geometry*, p. 99. The reference is to Lotze's *Metaphysik*, Bk. II, chap. ii.

Mr. Russell makes against Lotze is this: That observations such as Lotze contemplates would have to be due to departures from Euclidean straightness, hitherto unknown, on the part of stellar light-rays. Now if, as Lotze holds, some physical explanation is possible in such a case, the converse must also hold: it must be possible to explain the present phenomena by supposing ether refractive and space non-Euclidean. If every conceivable behaviour of light-rays can be explained, within Euclid, by physical causes, it must also be possible, by a suitable choice of hypothetical physical causes, to explain the actual phenomena as belonging to a non-Euclidean space.

The rejoinder is certainly effective (admitting, for the sake of the argument, that we have more than one notion of space), provided the object of Lotze's contention is that against which Mr. Russell argues. But even if this is not a matter of doubt, it is at least plain that Lotze had in mind the metageometer's standpoint that the several systems of geometry are all logically possible, that we cannot decide which of them is objectively valid from *a priori* considerations, but only, if ever, from actual measurements. Now whatever may have been the object of Lotze's argument, his actual contention is that all such measurements are irrelevant; in other words, the question as to the 'properties' of space can never be by this means indubitably settled; and obviously Mr. Russell's rejoinder is an excellent argument in support of this view.

It is one of the many 'little ironies' of the metageometrical position that while this view is, in substance, that to which Mr. Poincaré is led in the fifth chapter of *La Science et l'Hypothèse*, he nevertheless insists, as we have seen, that the question: Is Euclidean geometry true? (where 'true' can only mean 'objectively valid') has no meaning at all. These two views seem to be irreconcilable. If it is correctly argued that fresh experiences can never invalidate Euclidean geometry,[1] how can it at the same time be asserted that to ask whether Euclidean geometry is true or not has no meaning whatever? To assert that a question is devoid of meaning, and yet to give it a significant answer, seems to imply confusion of thought.

Returning now to the Euclidean standpoint: In applying

[1] 'La géométrie euclidienne n'a donc rien à craindre d'expériences nouvelles.' *Op. cit.*, p. 93.

Euclid's geometry to the results of astronomical observation we are said to assume that light is propagated in straight lines, the word 'straight' being used in its ordinary sense. It is, of course, a condition of our assuming the propagation of light to be rectilinear, that we can conceive light to be propagated in lines which are not straight, otherwise the assumption would be an empty one; and accordingly, as Mr. Russell observes, such measurements as Lotze supposes would have to be due to a departure from Euclidean straightness, to curvature (in the ordinary sense) of light-rays. But we clearly cannot leave the domain of pure geometry for that of its physical application without regard to physical theories—in this case the physical theory of light. We then perceive that although for ordinary purposes it is convenient to speak of the rectilinear propagation of light and of possible departures from rectilinear propagation, these expressions become not a little ambiguous in such a discussion as this. They do so, at least, in relation to the accepted theory that space is occupied by something which is called ether, and that a body which emits light creates, in this medium, some sort of disturbance which is propagated through it from the body as a centre. The assumption is that this disturbance progresses, from the body outward, uniformly in all directions, so that it reaches points equidistant from the body in the same time. In other words, neglecting the corrections for aberration, atmospheric refraction, &c., the assumption is that the axis or line of vision, which is normal to the advancing disturbance, is the direction in which the body lies. If we take into consideration the effect of the earth's atmosphere, we say that the light-rays are refracted or bent, which is merely a figurative way of saying that the wave-front of the disturbance is deformed in its passage through the atmosphere, so that, in general, the axis of vision does not coincide with the direction of the body: the body is not seen in the direction in which it lies.

Supposing, then, such cases established as Lotze imagines—say that the sum of the angles of stellar triangles had been found to be sensibly greater than two right angles, all the known corrections having been applied. The Euclidean pure and simple would have attributed this to the corrected directions of vision not coinciding with the directions of the star from the observer, which is the same thing as to attribute the result to some physical cause, e.g. an unknown obstruction to the uniform propagation of light. The metageometer, on the other hand, would argue that the cause need not necessarily be physical, that the result might quite well be consistent with a uniform propagation of light, that the corrected directions of vision did coincide with the directions of the star from the observer; only these lines would not be of the shape which Euclid calls straight, they would be the straight lines of Riemann's geometry, which return into themselves.[1] But when the alternative case is put in this way, there arises the question: To what notion does the term 'direction' refer us in connexion with non-Euclidean spaces? If we suppose a point to travel along a geodesic in a non-Euclidean space, direction must be defined in such a way that this point either does or does not change its direction in the process. If the term is defined so that the point does not change its direction, then there is a clear sense in which we can speak of direction of vision, direction of the star. But if the term is defined so that the point does change its direction, then we must at least be able to conceive motion without change of direction, for we clearly cannot conceive the one without the other. On this alternative supposition the conceiving of a non-Euclidean space involves either the notion of a path or line which, not being a geodesic, is not straight or, with respect to the geodesic, is curved; but along which nevertheless motion is without change of direction; or, again, since this appears to be an absurdity, the conceiving of a non-Euclidean space involves, as a necessity, the notion of some line which is neither a geodesic nor a non-geodesic of that space. But that the conceiving of a non-Euclidean space should necessarily involve that of an element foreign to it seems, once more, to be an absurdity.

Yet it is this alternative supposition, which leads to these apparent absurdities, to which we find ourselves committed, as is evident at the outset of Riemann's construction of the conception of the measure of curvature of space. He begins his explanation as follows:

'For this purpose let us imagine that from any given point the system of shortest lines going out from it is constructed; the

[1] The line of vision in any space is, at least according to Helmholtz, the 'straight' line of that space. See *Lectures on Scientific Subjects*, second series, p. 62.

position of an arbitrary point may then be determined by the initial direction of the geodesic in which it lies, and by its distance measured along that line from the origin.'[1]

Similarly, of course, 'a geodesic is completely determined by one point and its direction at that point.'[2] Clearly, according to this, a point which moves in a geodesic line may change its direction as it moves, or else there is no meaning in the 'initial' direction of a (fixed) geodesic or in its direction 'at a point'. It seems clear from this that the metageometer uses the term 'direction' in its ordinary sense. Now with the ordinary notion of space, and in ordinary geometry, the development of the correlative and complementary notions of straightness and non-straightness of line is perfectly comprehensible; these notions, notwithstanding that the former is particular and the latter general, are mutually dependent and condition one another. With any alleged non-Euclidean notion of space, and in non-Euclidean geometry, the process is incomprehensible: the notions of straightness and non-straightness of line at once are and are not correlative and complementary, at once do and do not condition one another. They are complementary and condition one another in so far that a line must be conceived as either 'straight' or not 'straight'; but in so far as the 'straight' line is conceived as a line which changes its direction from point to point, this conception has no complement in, and is unconditioned by, that of non-'straightness' of line; what makes it intelligible is not to be found in the alleged conception of non-Euclidean space, but has to be borrowed from that of Euclidean space.

When we reflect that the *co-existence* of different kinds of space is inconceivable—since we cannot conceive the co-existence of attributes which, *ex hypothesi*, exclude one another—while we at the same time admit that different kinds of space are possible to conceive, then at once we see also that the self-sufficiency of each of these conceptions, its independence, as a rational and intelligible whole, from all the others, is a logical necessity. But if, into the alleged conception of a non-Euclidean space, we import, in order to render it intelligible, a characteristic feature of Euclidean space, this logical condition is violated. If, on the other hand, this logical condition is to be

[1] Clifford's translation. *Mathematical Papers*, p. 62.

[2] Russell, *Foundations of Geometry*, p. 19.

fulfilled, each conception of space must involve a notion of direction peculiar to itself and not comparable with the others, just as each of these conceptions of space involves a 'space-constant' or notion of measurement peculiar to itself and not comparable with the others (Mr. Russell's discussion of the relation between the different space-constants will be recalled). The metageometer, however, cannot afford to make this admission, for he would thus give the *coup de grâce* to his own doctrine, since the admission must involve abandonment of Riemann's application of the conception of measure of curvature to space, and hence also of the conceivability of different kinds of space.

CHAPTER XX

GENERAL SUMMARY

PART I. The object of this part is twofold. It is, in the first place, an endeavour to get a clear understanding of the way in which the invention and employment of symbolic imagery develop and facilitate the process of thought in the individual. The discussion of this problem led to the comparison between the representative image and the symbolic image as signs or embodiments of concepts, and to the conclusion, suggested by that comparison, that the symbolic image, just because it symbolizes instead of representing, is far more effective than the representative image for the reinstatement in memory, and hence ready availability in discursive thought, of the more general concepts; more effective also for the classification of concepts in the order of generality. In this respect, and in this sense, the symbolic image may be said to be effectively instrumental in enlarging the sphere of the individual's conceptual system. The same order of considerations led to the further conclusion that the ratiocinative process is enormously accelerated by the use of the symbolic image as a substitute sign, but that in this respect the difference between words and mathematical symbols is superficial; is of degree, not of kind.

In the second place, the object is to insist upon the necessity, where considerations of an epistemological character are paramount, of always keeping a firm hold of the distinction between the process of thought and that of its symbolization, a distinction which the perpetual use of the symbol as substitute sign is for ever tending to thrust into the background. What has been called the vicious reaction of words upon the process of thought is, in substance, what I mean by the tendency to mysticism; and this tendency is very largely, if not wholly, due to letting this distinction go unrecognized; or, again, to the supposition that it is of no importance, which, indeed, is the case so long as the questions debated are purely logical, involve no epistemological problems.

It is hardly necessary to point out that a fairly adequate grasp of the general relations between the thought-process and the symbolic process, of the way in which the symbol operates as an instrument in the former process, is a prerequisite for the effective treatment of special or obscure cases of these relations or opera tions, which otherwise are likely to receive superficial or even altogether erroneous treatment.

PART II. The main conclusions reached in chap. iv are: (1) That the general conception of number is that of identity in aggregation. (2) That number-concepts are interdependent in their formation, the awareness of identity in aggregation necessarily involving that of difference, and of the least difference, in aggregation; and that consequently no effective definition of the name or symbol of a number can be given which does not involve the name or symbol of some other number, so that the name of one number at least must first be learnt by some means other than definition, in order that other numerical names or symbols may be understood by definition. (3) That the process of establishing identities or differences of aggregation being a process which in general involves attention to its order, creates that close association between the concepts of number and order which issues in the artifice characteristic of the invention of names for numbers as contrasted with the invention of other names, viz. their serial association in memory. Number-symbols thus necessarily become symbols of order as well as of identity in aggregation.

With reference to the universal custom of considering algebraic symbols indifferently as symbols of number or of quantity—a custom which Stallo appears to condemn at least by implication—we remarked that there is no real ground of objection, the numerical unit being indistinguishable save in name from the unit of purely abstract quantity.

Passing to the enigma involved in the Imaginary Quantities of Algebra and the Imaginary Loci of Geometry, and to the treatment of this problem first by Cayley and again by Mr. Whitehead, we noticed the difference of Cayley's attitude towards the Imaginary Quantities of Algebra on the one hand and the Imaginary Loci of Geometry on the other. He appears to have been satisfied that the development of algebraic symbolism does in reality lead to a more general and more fundamental notion of quantity, Imaginary Quantity, which includes the ordinary and less general notion 'Real' Quantity. But with regard to the Imaginary Loci of Geometry he was evidently not satisfied, since he ends by asking the very question which, ostensibly, he set out to answer, and invites further discussion of the subject. The most curious feature of his discussion is that, so far at least as the purely geometrical (i.e. non-analytical) side of the question is concerned, he himself appears to supply the answer without perceiving that the answer it is.

The view advocated by Mr. Whitehead, which seems to be that now generally entertained, is radically different from Cayley's. There is no question of any real extension of the conception of quantity, and the solution of the difficulty consists in conceiving Algebra as an independent science dealing with the relations of certain marks conditioned by the observance of certain conventional laws. The objection to this view, briefly stated, is this: That it consists in the substitution of one difficulty for another; that the more carefully we examine the arguments by which it is supported, the more clear it becomes that we have here a solution more ingeniously constructed than, but in principle no different from, that proposed by Boole, who solves the question by begging it.

In the two chapters which follow, the argument is designed to show:

(1) That the supposition or belief entertained by some mathematicians, and still commonly expressed in textbooks of Algebra, that the development of algebraic symbolism leads to a new and more fundamental, or more extended, notion of quantity than that with which Algebra starts, is an illusion which appears as the culminating point of, and for which the way is prepared by, the tendency to mystical explanation which marks the commonly received exposition of the conceptions and symbolism of Algebra in its most elementary phase. But the illusion is not wholly the result of the cumulative effects of this tendency; it is also in part due to an inconsistency, either unnoticed, or ignored as unimportant, in the logical development of the symbolic system itself.

(2) That if we reject the mystical implications involved in the orthodox exposition of elementary Algebra, steadfastly holding

to a rational view of its development ; and if at the same time we recognize the said inconsistency, admitting it for the sake of the formal symmetry which it confers on the system, then the conditions under which enigmatical questions suggest themselves entirely disappear. As Mr. Whitehead says, the difficulty vanishes—but Algebra does not vanish with it.

With regard to the geometrical interpretation of these so-called symbols of imaginary quantity, which, in Algebra pure and simple, are merely constituent parts of real symbolic contexts, and which cannot, in isolation from these contexts, logically be regarded as symbols, we have to distinguish between interpretation in the literal sense, and interpretation which, while it involves paradox in expression, does not involve paradox in the notions expressed ; provided these notions (so to call them) are not mystical—in other words, provided they are not illusions resulting from a vicious reaction of the mode of expression upon the judgement of the reasoner. The validity of these paradoxical or non-literal expressions (whether, as in pure geometry, the medium of expression is ordinary language or, as in analytical geometry, it is the symbolism of algebra) depends upon whether they do or do not denote real geometrical properties or relations. The test, then, is very simple ; for if a real geometrical property or relation is indicated, this can always be described in a literal manner, in phraseology which is not paradoxical.

PART III. The term Geometry has been employed here in much the same sense as Clifford uses it, namely, as denoting the science of shape, size, and distance. In this definition, however, the term ' distance ', if not the term ' size ', seems superfluous, the notion itself of distance or length being involved in that of linear shape. Clifford, it is true, speaks of geometry as being a physical science. But in this I am unable to follow him, for certainly in the general conception of the physical is contained that of interaction ; and by common consent this notion is quite foreign to geometry, though not to mensuration.

In Geometry, as in every other domain of thought, we can have particular notions only in virtue of having general ones, and vice versa : the notion of some particular linear shape, of some particular surface shape, is conditional upon that of linear shape in general, of surface shape in general. It is, then, manifestly erroneous to say that we can constitute geometry without these general notions. There is no conflict whatever between the fact that we require these notions in order to constitute geometry, and the fact that the process of geometry largely consists in 'defining' or determining these general notions in terms of particular notions.

The conception of shape is rooted in the experience of motion and the conception of motion itself involves that of direction. Thus the latter concept is a condition, but not necessarily a pre-condition, of the general conception of linear shape. This amounts to saying that straightness and curvature of line, permanence and change of direction in the continuous motion of a point are different ways of expressing essentially the same geometrical notion. The simplicity of the straight line resides in this uniformity of direction from point to point, which is obviously what constitutes it the inevitable standard of reference for linear shape, since change of any kind can be estimated only by reference to something given or assumed as changeless.

The notion of the straight line is not *a priori*; it is abstracted from experience, and is an absolutely definite notion, notwithstanding that we may admit the non-existence of straight edges or of motion in unchanging direction.

It is said that ordinary geometry is founded upon the assumption, among others, that the 'Euclidean' straight line exists; but we were unable to assign any definite meaning to this statement, either with regard to the straight line or to any other of the simple geometrical entities to which Euclid refers in the preliminary propositions of the *Elements*. Attention was drawn to the indiscriminate use of the term 'axiom' by various authorities, and to the diversity of views which prevails as to the nature of these propositions. For the sake of clearness in discussion, the term 'axiom' was confined to the type of proposition exemplified by Euclid's Axiom 1—that is, a statement of data and of a conclusion which follows at once necessarily and immediately from a synthesis of the data. Omitting the so-called axiom of parallels, and the so-called assumption of Free Mobility which Euclid uses (so it is said), an analysis of the fundamental propositions included among the preliminary propositions led to the conclusion that there are two axioms of magnitude, and that all the other fundamental propositions are definitions of geometrical abstractions.

Euclid's 12th Axiom, the axiom of parallels, is a proposition which is complex not only in expression but also from a logical point of view. It can be broken up logically into two propositions, the second one of which is geometrically equivalent to the converse of the first. These converse propositions I hold to be axioms in the sense already defined. The process of thought which they formulate is identical with the process of thought which the axioms of magnitude formulate. They are axioms of direction which Euclid ought to have given us in place of his 12th Axiom, which is, indeed, as Cayley says, self-evident—like those propositions the proof of which the Epicureans derided—but is not axiomatic.

Euclid, according to the modern view, makes use in the *Elements* of an assumption which he omits to state, viz. that figures can be freely moved without change of shape or size. It was contended, however, that Euclid with perfect consistency might have repudiated this interpretation of his procedure in demonstration, provided the authorities on Euclid's text are right in their opinion that the so-called Axiom 8 was added by later editors. However this may be, whatever Euclid might have had to say on the subject, an examination of the analyses of this supposed assumption, given respectively by the late Professor Clifford and by Mr. Bertrand Russell, leads to the conclusion that this assumption is relevant only to problems of mensuration. In geometry, where we make abstraction of the conceptions of cause and effect, physical interaction, 'non-geometrical circumstances'—to use Clifford's expression—the statement that figures can be moved freely without change of shape or size is, as an assumption, meaningless; it merely defines a constituent of the notion of metrical relation, i.e. 'identical' equality or congruence.

There are thus, in geometry, four axioms (or eight, if the point of view taken in the note to chap. xiii is admitted); one pair relevant to the notion of magnitude, the other to the notion of direction—the two most fundamental notions of geometry. In each case the pair is a pair of converse propositions. The other premisses of geometry are propositions which define the fundamental geometrical abstractions, or are these abstractions presented as ' indefinables '. Assumption, in the ordinary sense of the word, has no relevance to geometry

With chap. xvi we pass from what may be called the positive side of the argument respecting Euclid to the negative side. No matter how convincing a case has been, or might be, made for the apodeictic certainty of ordinary geometry, it must necessarily fall to pieces if, as is asserted to be the case, other systems of geometry are really conceivable, provided of course that the term ' geometry ' is throughout employed in the same sense. The question is, then, are other systems of geometry really conceivable ?

Comparing what are called the mutually exclusive planimetries of Euclid and of Lobatschewsky, it was pointed out that the latter only contradicts the conclusion affirmed in Euclid's 12th Axiom provided his own affirmation is relevant to the data of Euclid's proposition, that is, relevant to the surface and line which Euclid calls respectively plane and straight, and that this appears to have been Lobatschewsky's intention. In that case, admitting the premiss substituted by Lobatschewsky for Euclid's —which, save as a mere exercise in logic, we can only do if we consider Euclid's 12th Axiom to be not apodeictic—Lobatschewsky establishes an alternative planimetrical system to Euclid's, where ' plane ' has the same meaning in the two systems. But if it cannot be admitted as possible that we can have two mutually exclusive sets of metrical relations for Euclid's plane, this is equivalent to admitting that Euclid's 12th Axiom is apodeictic ; and in that case Lobatschewsky establishes a self-consistent metrical system which rests upon a false premiss. On the other hand, if it is held that Lobatschewsky's premiss is relevant to some other surface than that which Euclid calls the plane, then Lobatschewsky does not really contradict Euclid, he does not establish a different system of geometry, but he gives us the geometry of a surface of constant negative measure of curvature. We can escape from this argument only by falling back upon the meaningless statement that Euclid ' assumes the existence ' of the surface of zero curvature, and that Lobatschewsky rejects this ' assumption '. But the question assumes a totally different aspect when it is alleged that these two geometries, and subsequently the third geometry of Riemann, arise respectively from three different conceptions of space, and that we have thus to distinguish between the two notions: (1) the sphere and the pseudo-sphere in 'Euclidean' space, (2) the plane in each of the two 'non-Euclidean' spaces.

The significance of the metageometrical doctrine thus depends upon whether we can in fact conceive different kinds of space. But neither in the popular expositions of Helmholtz and Clifford, nor in the academical dissertation of Riemann, is there any evidence that this is really possible, while on the other hand there is in all of them evidence of real confusions of thought. Helmholtz's explanation confounds what may briefly be described as the logical treatment of an allegory, with a development of analogy in conception; and Clifford's rests upon what appears to be simply a false analogy. Riemann, as we saw, and as seems to be admitted by at least some metageometers, fails to construct a general conception of externality which includes as a particular case of it the ordinary notion of space. The conception which he calls that of an *n*-fold extended magnitude, and to which his purely mathematical argument is relevant, is the conception of a general system of purely abstract quantitative relations, or of a series of particular systems of such relations. He calls these systems of quantitative relations possible systems of measure-relations of space. But this is permissible only if a general conception of externality has been constructed, otherwise the question is begged, since this general conception is the necessary condition of the systems of abstract quantitative relations being conceived as systems of spatial measure relations. We do not escape this conclusion by merely talking about analytical conceptions of space, or by referring to non-Euclidean conceptions of space as being analytical constructions, for until that in which Riemann failed is accomplished, these are and remain question-begging phrases.

In Cayley's Theory of Distance we have a doctrine which is also metageometrical or non-Euclidean, though in a completely new sense. Here there is no longer any question of conceiving different kinds of space; but, nominally, of conceiving different systems of metrical relations in space. I say 'nominally' because the systems of relations are metrical in name, while in conception they are descriptive. We do not in fact conceive different systems of metrical relations, but, according to the choice of the fundamental figure which Cayley calls the Absolute, we have different systems of descriptive equivalence. The supposition that we conceive different kinds of distance is an illusion engendered by an unrealized reaction of symbolic forms, verbal or algebraic, on the thinker's judgement of his own process of thought. It leads him to say, e.g., that the ordinary notion of distance is included, as a special case, in the generalized notion of distance, while the real fact is that this generalized notion of distance is generalized, not from the ordinary notion of distance, but from a notion of projective relation.

Finally, in discussing the relation of mensuration to metrical geometry, in the literal sense of the term 'metrical', it was contended that the view of this relation commonly entertained by metageometers is indefensible. For if Mr. Russell's criticism of Lotze's argument for the apodeictic certainty of ordinary geometry is valid, it is no less so when turned against the belief that the objective validity of ordinary geometry might be disproved by greater precision or larger amplitude of astronomical observation. The further discussion of this relation, from the points of view respectively of the Euclidean and the Metageometer, led to the question of the relevance of the notion of direction to that of non-Euclidean spaces, and to the conclusion that, were such spaces really conceived, the ordinary notion of direction could not find a place in these conceptions of space, and would logically be irrelevant to them; their construction would involve non-Euclidean notions of direction. Thus the fact that the ordinary notion of direction is employed in the construction of these alleged non-Euclidean conceptions of space would alone be sufficient to warrant the rejection of the doctrine that different kinds of space are conceivable.

INDEX

(*See also Table of Contents*)

Oxford: Horace Hart, Printer to the University

Printed in Dunstable, United Kingdom

91219923R00157